Contents

Tables

Figures

Contributors

Chapter 1

Ryszard Borowiecki is a full professor at WSB University in Dąbrowa Górnicza. From 2002 to 2008 he served as a rector at Krakow University of Economics. He is also a researcher with 54 years of experience and is the author, coauthor, and/or scientific editor of more than 600 works in the fields of corporate management, economic analysis and diagnostics, and corporate development.

Zbigniew Olesiński is a full professor at Vistula University in Warsaw, who specializes in interorganizational management, inter-organizational networks, clusters, and globalization problems. He has participated in or managed 11 research programs and 13 EU research grants. He is also the author or coauthor of 206 publications, including 14 books.

Agnieszka Rzepka is a professor at the Lublin University of Technology. She specializes in inter-organizational management, intellectual capital, self-management, and SMEs. Her current research is focused on teal organizations and Industry 4.0. She has participated in or managed three research programs and 10 EU research grants. She is the author of 135 papers, including 7 proprietary and collective monographs.

Chapter 2

Sabina Kauf is a full professor of Management in the Institute of Management and Quality Sciences at the University of Opole, Poland. Her research interests focus on the interface between marketing and logistics and the use of modern technologies in supply chain management and smart city. Her research has been published in a wide range of peer-reviewed journals. She has also been an investigator of Polish research projects.

Iwona Pisz is an associate professor of Management in the Institute of Management and Quality Sciences at the University of Opole, Poland. Her main research field covers uncertainty modeling in project management, production management, logistics, and supply chains. She has published widely on these themes in academic journals and books and has presented at international conferences and high-level meetings. She is an investigator of research projects financed by the European Union.

Chapter 3

Sandra Grabowska, Dr. Eng., is an assistant professor at the Department of Production Engineering (Silesian University of Technology). She is the author and coauthor of nearly 60 scientific publications. Her research focuses on Business Models, modern forms of management, and Industry 4.0. She is also active in managerial associations and as a consultant supporting enterprise improvement.

Sebastian Saniuk, Dr. Eng., is employed as a professor at the University of Zielona Góra. Since 2015 he is a member of the Committee of Production Engineering of the Polish Academy of Sciences. His research focuses on production management, Industry 4.0. The results of his research have been presented in over 220 scientific publications: journals, monographs, and conferences.

Chapter 4

Mirosław Moroz is an associate professor of Management at the Wroclaw University of Economics. His research is mainly focused on e-commerce, e-marketing, and Industry 4.0. He has published more than 90 papers in peer-reviewed journals.

Chapter 5

Michał Baran is a post-doc lecturer of the Institute of Economics, Finance and Management at Jagiellonian University, Kraków, Poland. His work analyses information flow from the strategic perspective. He is the author or coauthor of about 70 scientific papers and books.

Chapter 6

Radosław Luft, Ph.D., is a lecturer at the Kazimierz Pulaski University of Technology and Humanities, Faculty of Economics and Finance in

Radom, Poland. His scientific interests relate to the impact of integrated information systems on enterprises development. He conducted research on the impact of IT systems on the enterprise's competitiveness in the SME sector. He is an author of many national and foreign publications.

Anna Wolak-Tuzimek is an associate professor at the Kazimierz Pulaski University of Technology and Humanities in Radom, Poland. She is the head of the Department of Economics at the Faculty of Economics and Finance. Her research centers around conditions of enterprise development. She studies the impact of Corporate Social Responsibility on enterprise competitiveness. She has authored and coauthored more than 110 research publications, including 8 monographs published in Poland and abroad.

Chapter 7

Joanna Martusewicz is an assistant professor of Management at the Wroclaw University of Economics and Business, Poland. The main area of research interests focuses on leadership, designing management systems, and models of business excellence. She is an author of several dozen articles, textbooks, and scientific monographs and a member of several research grants. She is a licensed EFQM trainer and assessor and a member of the Committee of the Polish Quality Award.

Karol Szewczyk is a head of Manufacturing Engineering Department, Robert Bosch Hungary – Miskolc plant. He is an MBA graduate, Industry 4.0 enthusiast, and works in the area of production and quality on many managerial international positions.

Arkadiusz Wierzbic is an associate professor of Management at the Wroclaw University of Economics and Business, Wroclaw, Poland. His work is dedicated mostly to Quality Management. In recent years, his research is focused on excellence models. He is an auditor of quality management systems and an assessor of the EFQM Global Award.

Chapter 8

Anna Barwińska-Małajowicz is an associate professor of Economics at the University of Rzeszów, Poland. Head of the Department of Economics and International Economic Relations. Her works analyze international economic relations, entrepreneurship, and the labor market. Her research has been published in a wide range of peer-reviewed journals. She is an author of nearly 180 scientific publications and an expert and principal researcher on a number of Polish and European projects (labor market).

Patrycja Żegleń is a Doctor of Economics at the University of Rzeszów, College of Social Sciences, Institute of Economics and Finance, Poland. She graduated from the Warsaw School of Economics. Her scientific interests focus on public–private partnership, entrepreneurship, and economics of tourism. She is an author of over 100 scientific publications. She took part in several European projects (both scientific and didactic).

Chapter 9

Tomasz Derlecki is a deputy IT manager at Walstead CE, Poland. He was responsible for implementing autonomous mobile robots in this company. He is also a teacher at WSEI College of Economics and Computer Science in Krakow. His area of research interest includes Industry 4.0, data analysis, Big data, data mining, and machine learning.

Jerzy Duda is a professor in the Department of Applied Computer Science at AGH University of Science and Technology, Poland. He is a contractor of many projects in the field of ICT and production management; currently, he is the head of the research team in an Industry 4.0 project and a member of IEEE and EU/ME. His research interests include AI in production management, heuristic algorithms, and APS systems.

Robert Goncerz is the chief of Industrial Automation and Robotics Solutions. He is a practitioner, builds innovative management and automation systems for production processes, starting from the concept of a solution, through research, analysis, and design, to subsequent implementation in a production plant. His solutions were presented in many industry magazines. He also participated in research projects co-financed by the EU.

Daniel Kubek is an assistant professor at the Cracow University of Technology, Poland. His main areas of research are modeling and simulation of transportation and logistics systems, including optimization techniques. He is currently working on a project concerning a cyber-physical logistic platform in a printing company.

Radosław Puka is a Ph.D. in the engineering of production. He works as an assistant professor at AGH University of Science and Technology, Poland. In his research work, he focuses on optimization, scheduling, and the use of artificial intelligence. He participated in many projects in which he had the opportunity to verify scientifically developed methods in business.

Katarzyna Rybicka is Vice-President responsible for solution delivery and R&D at Walstead CE. Over the years, she has played managerial roles in production, distribution, and financial sector as Strategy, Innovation, and Project Manager, involved in the biggest projects related to the transformation of the financial sector in Poland, and also experienced in FMCG, winner of the Prize of Innovation in Sales in 2013 in Polish National Sales Awards contest.

Iwona Skalna is a professor of Computer Science at the AGH University of Science and Technology, Poland. Her research work mainly focuses on the modeling of epistemic uncertainty. Other interest areas include data mining and artificial intelligence. Her research has been published in a wide range of peer-reviewed journals. She participated in several research and innovation projects.

Paweł Więcek is a Ph.D. Eng. in Civil Engineering and Transportation. He works in the Transport Systems Institute at the Cracow University of Technology. His research focuses on logistic systems optimization, computer simulations in logistics, and the application of artificial intelligence tools in logistic processes control and management. He is an author and coauthor of many publications in peer-reviewed journals and a participant in many research projects.

Chapter 10

Mariusz Trela is an assistant professor at the Faculty of Management at the AGH University of Science and Technology. His works concern economic and environmental aspects of transport, with particular emphasis on new technologies in road transport. His research has been published in a wide range of peer-reviewed journals. He participated in many projects related to road transport.

Chapter 11

Sandra Grabowska, Dr. Eng., is an assistant professor at the Department of Production Engineering (Silesian University of Technology). She is the author and coauthor of nearly 60 scientific publications. Her research focuses on Business Models, modern forms of management and Industry 4.0. She is also active in managerial associations and as a consultant supporting enterprise improvement.

Sebastian Saniuk, Dr. Eng., is employed as a professor at the University of Zielona Góra. Since 2015, he is a member of the Committee of Production Engineering of the Polish Academy of Sciences. His research focuses on production management, Industry 4.0. The results of his research have been presented in over 220 scientific publications: journals, monographs, and conferences.

Chapter 12

Waldemar Jędrzejczyk is an associate professor of Management at the Czestochowa University of Technology, Poland. His work analyses the issue of competencies, both currently desirable and prospective, in relation to sectors and types of organizations as well as professions and key positions in organizations. His research has been published in a wide range of peer-reviewed journals. He has also been a principal investigator on several international projects.

Chapter 13

Jacek Jakieła, Ph.D. Eng., is a university professor at the Department of Computer Science (Rzeszow University of Technology, Poland). He is the author of several scientific papers and two textbooks on business computing as well as a coauthor of international projects regarding the application of the Design Thinking framework and creativity-oriented methods for students' professional development (e.g., the BEAST Project).

Joanna Świętoniowska, Ph.D., is an assistant professor in the Department of Management at the University of Information Technology and Management in Rzeszow, Poland. Her research and teaching expertise includes project management, especially the project management maturity of organizations as well as entrepreneurship and innovations. She is the project manager of the BE Aware STudent [BEAST] project.

Joanna Wójcik, Ph.D., is an assistant professor in the Department of Cognitive Science and Mathematical Modeling at the University of Information Technology and Management in Rzeszow, Poland. Her main research focus is the domain of new technologies, e-learning, university management, business models, and agile methodologies. She is the e-learning officer at the University of Information Technology and Management in Rzeszow.

Chapter 14

Damian Kocot is a Ph.D. in Economic Sciences, an assistant professor at the University of Economics in Katowice (Poland), works at the Department of Economic Informatics, and an author of scientific articles in the field of management and economics.

Magdalena Maciaszczyk is an assistant professor at the Faculty of Management, Lublin University of Technology, where she held the position of Deputy Dean for Students' Affairs. Her research interests include marketing, marketing communication, advertising, or consumer behavior. She is an author and a coauthor of over 30 articles and author/editor of a few books. She is also currently working on ethnocentrism and presumption.

Chapter 15

Barbara Siuta-Tokarska is a professor in the Department of Economics and Organization of Enterprises Development at the Cracow University of Economics. She is an author and a coauthor of more than 150 publications on sustainable and permanent development of economy and enterprises. She authored six books on the development of small and medium enterprises. Her research area also includes issues of social challenges and dilemmas of the New Economy in the 21st century.

Agnieszka Thier, Ph.D., is an assistant professor in the Institute of Economics and Corporate Organization at the Cracow University of Economics. She is an author and a coauthor of 70 publications on corporate management, structural changes in power generation and gas industries, water management, environment management, sustainable development, social entrepreneurship, the digital revolution and its consequences, and family business management in the contemporary economy.

Preface

For many years, there was a clear distinguishing line between physical systems used in the industry and digital systems. During the past few years, this line has blurred and the term "cyber-physical systems" (CSP) has emerged, which is used to denote systems in which physical objects, labor resources, and global data networks are intertwined through information technologies (IT). In this way, they enable industrial operations that are controlled and driven by IT systems, which give a rise to the concept called "Digitizing Industries." What is more, these processes are developed in both global and local dimensions.

The digitization of industries, which is the cornerstones of the fourth industrial revolution (Industry 4.0), offers new possibilities for optimization and automation of all business processes; so, it is not surprising that it has been the subject of interest to the researchers representing various areas of science such as economics, management, sociology, technology, and computer science (Machado & Davin, 2020). Papers published in the literature, alongside strictly technical ones, refer to the impact of modern technologies on the functioning of enterprises, the economy and society among others. The most frequent topics include optimization of logistics and supply chains, autonomous machines and vehicles, robotics, additive production, Internet of Things (IoT) (De Saulles, 2016), and cloud computing (Alcácer & Cruz-Machado, 2016).

With this book, the authors want to share their own experience and research related to the above-mentioned areas. They give new insights and present recommendations resulting from their own research and analyzes as well as experiences in creating and implementing solutions for enterprises. This book addresses the following four basic issues related to the impact of Industry 4.0 on the functioning of enterprises from a glocal perspective:

- *Personnel management* which is very important in organizations, especially in the face of the fact that machines are increasingly replacing people in their workplaces; new problems arise in this area that require completely new approaches.

- *Marketing* which increasingly requires the use of electronic forms of communication and various automated tools adapted to the individual needs of consumers.
- *Technical solutions* which have a large impact on the increase in the efficiency and flexibility of enterprises; they allow, among others to shorten production time, increase efficiency, and minimize the risk of producing defective products.
- *IT systems*, including the use of advanced communication technologies and AI algorithms, which are used to automatically control the production process, as well as support in solving decision problems at both operational and strategic levels.

Case studies presented in individual chapters show the opportunities offered by the 4.0 industry concept for the functioning and development of enterprises, sectors of the economy, or countries. Authors present benefits and threats resulting from the changes brought by computerization, digitization, robotization, and introduction of new management models using artificial intelligence. The critical review of the world and national literature on the subject is provided as well, both for comparison purposes as well to show the importance of the topics discussed for the future of the national or international economy. The last chapter refers to a new trend called Industry 5.0, which returns to the role of human in improving the efficiency of production processes through appropriate interaction of human workers with machines.

This book is addressed to business owners, managers of global corporations and smaller enterprises, engineers, scientists, future university graduates, and all those interested in the possibilities offered by the concept of Industry 4.0.

Jerzy Duda and Aleksandra Gąsior

References

Alcácer, V., & Cruz-Machado, V. (2016). Scanning the industry 4.0: A literature review on technologies for manufacturing systems. *Engineering Science and Technology, an International Journal*, 22(3).

De Saulles, M. (2016). *The internet of things and business.* Routledge Focus.

Machado, C., & Davin, P. (2020). *Industry 4.0 challenges, trends, and solutions in management and engineering.* CRC Press and Taylor & Francis Group.

1 Evolution of Organization Management in Light of Economy 4.0 Challenges

Ryszard Borowiecki, Zbigniew Olesiński, and Agnieszka Rzepka

Introduction

The aim of this chapter is to propose ways of management evolution in the emerging Economy 4.0. The authors of this chapter have assumed that, under such conditions, the evolution of organizations is particularly important in the creation of a so-called "Teal organization." Such organizations, which are characterized by the customizable self-management of its members (employees) in decentralized structures (Holacracy), are developing faster under those conditions that support agility, inter-organizational collaboration, and the intensive creation of the soft factors of an organization's management due to the development of computerization and innovation.

This relationship is demonstrated later through an analysis of quantitative data from previous surveys conducted by the chapter's authors at hundreds of organizations located in Poland and abroad. Immediately after conducting the research, we noticed a correlation between the research results and the formation of Teal organizations. Therefore, we intend to continue these surveys in the future to further our understanding of the impact of selected factors on the creation of Teal organizations. Prior to this, however, we strongly feel that there is a need to demonstrate our findings in this work.

Characteristics of Economy 4.0 in Poland

The 4.0 symbol is used to describe the Fourth Industrial Revolution, which is characterized by the trend toward automation and artificial intelligence. The shift is clearly observable in the modern era and comes after the First Industrial Revolution of coal and steel at the beginning of the 11th century, the Second Industrial Revolution at the turn of the 20th century (a revolution in chemistry, the automotive industry, and electricity), and the Third Industrial Revolution of the 1960s—the automation revolution.

DOI: 10.4324/9781003186373-1

Industry 4.0 was originally presented in Germany at the Hannover Messe in 2011 (Roblek et al., 2016, pp. 1–10); it is associated with cyber-physical systems (CPS), cloud computing (CC), the Internet of Things (IoT), and big data. Its main goal is to achieve accuracy and precision as well as a higher degree of automation (Thames & Schaefer, 2016, pp. 12–17). Due to the fact that it is a rather recent phenomenon, the Fourth Industrial Revolution has not been sufficiently described in the scientific literature. It is quite difficult to say exactly what form the aforementioned automatization will take (and even more so for artificial intelligence). The history and development of earlier industrial revolutions allow us to state that, according to propagators of this idea, it will be another great breakthrough in the history of humanity and, in part, the human activity understood as production.

Despite the rapid growth of automation in Poland, it must be clearly stated that automated machine work is generally more expensive than human work (as can be seen in the works of Schwabe, 2018, p. 65). Hence, hiring employees is preferred to introducing machines despite the fact that the cost of automated machines has been decreasing; it is currently more profitable to use automation instead of human laborers in harmful conditions (high temperatures, toxic fumes, etc.). The widening use of IT will significantly facilitate the functioning of modern organizations even though the use of the term "Artificial Intelligence" could be currently exaggerated at this time.

Undoubtedly, it is advisable to use the so-called Internet of Things (IoT) to an increasing extent (within the limits of economic viability, of course). These are electronic signaling devices that activate specific organizational processes (including production) according to the adopted programs. The world is on the cusp of rolling out an early fifth-generation IT revolution (symbolized by the phrase "5G"), which will enable much faster data circulation. This will also include specific relationships in the field of automation control (such as turning devices on and off).

In Poland, one should take the acceleration of automation development processes found in functioning organizations into account, including organizations that cooperate with companies from developed countries (the United States and Western European countries) and participate in globalization processes. It also seems quite possible that we will see the rise of a completely new type of organization—one that will use the advanced IT processes and advanced automation available in a given sector from the beginning (e.g., start-ups, IT, design, nanotechnology). It, therefore, seems purposeful to monitor these processes in the Polish economy and Polish society as well as to formulate postulates that will help stimulate the above processes.

In some circles, it is believed that the development of IT will lead to algorithms replacing increasingly large groups of employees (i.e., technologically advanced machines). It is believed that this leads to dataism (i.e., the inflow of a significant amount of data that requires ongoing processing, which can be carried out by devices using artificial intelligence):

> The work of processing data should therefore be entrusted to electronic algorithms, whose capacity far exceeds that of the human brain. In practice, this means that Dataists are skeptical about human knowledge and wisdom and prefer to put their trust in big data and computer algorithms (Harari, 2019a, p. 32).

Extensive caution should be used in all forecasts, however. In *21 Lessons for the 21st Century*, Y.N. Harari states that the changes will be more subtle in the foreseeable perspective of 20–30 years and that algorithms will probably not make as much progress as predicted in *Homo Deus*.

> I discussed the merger of infotech and biotech at length in my previous book, *Homo Deus*. But whereas that book focused on the long-term prospects—taking the perspective of centuries and even millennia—this book concentrates on the more immediate social, economic, and political crises. My interest here is less in the eventual creation of inorganic life and more to the thread of the welfare and to particular institutions such as the European Union (Harari, 2019b, p. 98).

Therefore, creating such safeguards before implementing specific innovative solutions seems to be an urgent need. Not only are legal safeguards defining testing procedures and methods of implementing new technical and technological solutions but also specific technical and technological solutions—control over specific elements of energy systems with the possibility of disconnecting certain elements from the power supply, the possibility of physical interference against a specific element of the system (which can tend to get out of control). It seems that contemporary societies have the appropriate tools to be able to oppose the negative relationships mentioned earlier.

Trends in Management Toward Industry 4.0

Management and quality as a science took shape in the 20th century. The assumption of this text is to show the ways this science developed from the beginning of the 20th century to the present day in order to describe, explain, normalize, and forecast management and quality procedures. A convenient starting point for such an analysis will be the indication of the following diagram by A. Koźmiński that characterizes systemic management.

This diagram allows us to indicate two main ways of current developments in the field of management and quality—social processes (upper part of the figure) as well as technical and technological processes, including IT and automation (lower part of the illustration).

The social process has existed in the science of management and quality since at least the 1930s. Rapid development in this area began in the early 1960s, when concepts such as the theory of organizational game and the theory of organizational balance appeared. In this context, the concept of a so-called Teal organization is understood herein as a result of continuing changes that, in the last several years, have found their way into emerging organizations whose paradigms are characterized by self-organization and self-management.

This concept stems from the previous reflections on the social, psychological, sociological, political, anthropological, and ethnographic conditions observed in the science of management and quality. This allows us to pose the main research question of whether the intensification of management processes occurs particularly effectively in the course of the development of automation processes and artificial intelligence through the development of self-management (particularly in Teal organizations). Answering the above question calls for an examination of the significance of the development of agile employees and agile organizations, development of inter-organizational collaboration, and development of conditions for the creation of the soft factors of an organization's management (such as knowledge, intellectual capital, trust, and intercultural conditions). Additionally, as these issues are found in the processes of shaping Teal organizations, there is a question of whether and to what extent they can influence the creation of conditions conducive to the intensification of innovation.

In the context of the main question and five auxiliary questions above, the following diagram can be formulated (Figure 1.1).

Innovation will be an important driving factor in the emerging Economy 4.0. The evolution of science to date indicates that properly prepared employee teams can, under favorable conditions, significantly and systematically propose new technical and technological solutions. Solutions that save the exploitation of previously known resources and improve environmental impact are of crucial importance. An example of such a solution could be obtaining electricity from wind or the sun, while electric motor vehicles are yet another.

Today, the development of IT processes (especially artificial intelligence) is particularly important. These solutions can pose threats that automated machines get out of human control (Harari, 2019a). However, using the appropriate managerial rigors and specific social and political security measures, this threat can be significantly reduced; hence, the role and significance of the above-mentioned managerial factors, such as agility, inter-organizational collaboration, or soft factors of an organization's

Figure 1.1 Model of factors accelerating the formation of Teal organizations
Source: authors' own research

management. It should be assumed that the science of management and quality will create new types of employee and manager behavior; these will have specific procedures to effectively reduce the risk of automated machines slipping out of control. It is also important to find mechanisms that will oppose the possible loss of control over some elements of artificial intelligence.

The Role of Agility in Teal Organizations

The transformations taking place under the conditions of both a changing environment and the diminishing role of globalization processes have increased the interest and directed the attention of scientists to the role of intellectual capital in creating and achieving market position by organizations. Under the conditions of growing competition, only organizations that aim to continuously improve their positions through the proper management of their intellectual and human capital stand a chance for development and survival. Along with their skills, knowledge, and competencies, employees are treated as the basic source of achieving a market advantage. In an environment based on competition, organizations that are agile and flexible and able to react quickly will grow.

Although many different definitions of "agility" and "agile manufacturing" exist in the literature, these terms are mainly understood as the "ability to quickly respond and adapt in response to continuous and unpredictable changes of competitive market environments" (Kidd,

1995, p. 31). The successful and fast response to changes requires that an agile organization is able to adapt all organizational elements such as goals, technology, structure, and people to unexpected changes.

In the current competitive environment, only those organizations that are agile and flexible are developing (above all, those that are able to react quickly). The ability to adapt to change becomes a strategic determinant of survival. Efficient and effective organizations are designed in such a way that employees can quickly adapt to market changes. The employees of these organizations are more focused on collaboration and take risks more easily to achieve the result. They also show a greater ability to act in an ever-changing environment in line with their personal potential; this is because they get the support they need from their leaders. Speaking of an agile organization, we must pay attention to its key components—the most important of which are speed and flexibility. In addition, the effective response to changes and uncertainty in the business environment is crucial today.

Organizations will and should differ from each other in a modern and dynamically changing business environment, mainly in relation to their potential and how it can be used to achieve and maintain agility. Today, organizations must implement the above features to survive in an environment of growing change. Organizations must respond to changing conditions and respond fairly quickly to emerging opportunities and threats. First of all, they should acquire the ability to assess the correlation of their resources as well as the ability to obtain them. According to the Laloux concept, the individual predispositions and advantages of each employee are most important in a company management model. Only a skillful approach of combining process and operational management with vision can support an organization in its functioning in a turbulent environment and the implementation of Teal management.

Role and Importance of Inter-Organizational Collaboration

Inter-organizational relationships between independent entities develop along with intellectual capital. These relationships are currently considered to be one of the key development trends in contemporary organization-creation mechanisms. A variety of concepts appear in management sciences and organizational theory, emphasizing the growing importance of inter-organizational cooperation. Networking has emerged as a key category in management science, and practical examples of inter-organizational relationships have become a ubiquitous phenomenon in economic reality.

Co-existing with other market players, organizations enter into various relationships and with one another (Oliver, 1990). In a network of interactions, it is possible to isolate a variety of attitudes that are

Table 1.2 Collaboration with other organizations compared to respondent's profile, *N*=202

(answers on a scale of 1 to 7, where 1 stands for "never" and 7 for "always")*

Company's characteristics		*No.*	*organizations*	*Consulting companies*	*Research centers*
Period of time in operation	Less than 1 year	7	2.43	1.71	1.14
	1-3 years	29	3.17	2.38	2.55
	4-7 years	33	3.06	2.67	1.88
	8 years or more	133	3.08	1.94*	1.68*
Type of business	Industry	60	3.07	2.40	2.03
	Trade	57	3.14	2.07	1.86
	Services	66	2.91	1.85	1.47
	Other	19	3.37	2.26	2.26*
Geographical range	Local	102	3.02	2.04	1.75
	Regional	58	2.88	2.00	1.76
	European	33	3.52	2.45	2.18
	Global	9	3.11	2.44	1.78
Number of employees	10 or fewer	49	3.33	2.16	1.78
	10–49	97	2.94	2.10	1.80
	50–249	29	3.10	2.41	2.03
	250 or more	27	3.00	1.74	1.74
Respondent's sex	Female	94	3.11	2.19	2.03
	Male	108	3.01	2.02	1.59*
Respondent's age	25 or younger	48	3.52	2.00	1.63
	26–35	65	2.60	2.15	1.83
	36–45	59	3.07	2.10	2.02
	45 or over	30	3.33	2.23	1.73
Years of respondent's professional experience	5 or fewer	81	3.17	2.17	1.84
	6–10	42	2.79	2.07	1.95
	11–15	28	2.96	2.25	1.89
	16–20	21	2.57	1.48	1.05
	20 or over	30	3.60	2.33	2.07
Respondent's position	CEO	25	3.08	2.84	1.92
	Middle management	33	3.61	2.76	2.27
	Low-level management	57	2.89	2.05	1.93
	Others	87	2.97	1.70**	1.55*
Total		202	3.06	2.11	1.82

*Statistically significant variables: *$p<0.05$; **$p<0.01$.
Source: Rzepka, 2018, p. 214

of the growing importance of intangible assets in management. The faster and faster pace of changes in an organization has also become significant, which in turn is the result of the impact of an unpredictable and turbulent environment. Success in managing inter-organizational relationships lies in taking the autonomy of the participants operating in a network into account and ensuring the coordination necessary for their proper cooperation. An important issue is also the layers of knowledge that, along with the abilities and competencies, decide on obtaining a strong position on the market by a modern organization. All of these prove helpful in increasing the innovativeness of an organization.

For these processes to take place correctly, it is necessary for all participants to be active, monitor their situation, and stimulate demands forced by contestation and disobedience. There will always be a temptation to seize public space by certain groups of people and organizations (hence, the need for activity, constant negotiation, and consensus). Therefore, soft factors play an essential role in creating inter-organizational space. This leads to its institutionalization, facilitates the conciliatory model of functioning, and creates inter-organizational space in the Teal organization model.

Innovation in Teal Organizations

The ability to constantly search for, to implement, and to disseminate innovation is understood as the innovation of an organization (Juchniewicz & Grzybowska, 2010, p. 102). This is the ability to effectively allocate resources to shape the optimal configuration of competitive advantages. Innovation defined in this way contains an element of efficiency and a factor of time; under such an influence, the shape of the optimal configuration of competitive advantages changes.

Innovation is treated as a practical transformation of ideas into new products, services, processes, systems, and social interactions. It creates new streams of value aimed at satisfying stakeholders and driving sustainable development. It creates jobs, increases the quality of life, and also promotes the sustainability of society. The innovation process does not only apply to "high technology" but has become present in all areas of the economy and society.

In the science of management, innovation is treated as an important asset of an organization, manifesting itself in the ability to introduce innovations. To a large extent, organizations undoubtedly owe their market success to innovation. Innovation remains an extremely important factor in building competitive advantage, but it is also vital in the implementation of new ideas and solutions. Its derivative result is the ability to create as well as implement and absorb innovations. Innovation is inseparably connected with an active involvement in innovative processes and undertaking efforts in this direction (Zakrzewska-Bielawska & Walecka, 2013, p. 317).

As with development, innovation is one of a company's inseparable assets. However, it is located in a group of features that do not result from the existence of specific relationships between the organization and its environment but are a requirement for conducting business in a market economy in such a way that determines the ability to function, survive, develop, and succeed in a difficult competitive reality (Nowacki, 2010, p. 28). Some elements of innovation systems (such as technology, innovation, and human capital) are factors that affect the competitiveness of organizations. Entities benefiting from innovation (i.e., organizations as well as scientific and research units) directly and indirectly shape competitiveness factors through their business or scientific activities. The economic collaboration with foreign entities undertaken by these entities is becoming an important channel for the exchange of new knowledge and innovation, so it is a very important factor of innovation and competitiveness (Glugiewicz & Gruchman, 1998, p. 158).

The authors' research mentioned earlier also included elements that are related to innovation. An important area concerned collaboration based on a common innovative product, service, technology, or organizational system.

Nowadays, employee participation is an attractive method and management tool at the same time, as it contributes to the increase of an organization's efficiency through creating the right climate for innovation, among other things (Borowiecki & Rojek, 2015, p. 108). People have certain formal qualifications as well as other skills (Elsbach & Kramer, 2015, p. 89). In this sense, innovative management is about enabling them to use their qualifications and skills to meet their interests—otherwise, it is the art of creating the conditions for fulfilling the desires of individual employees in connection with the function of the organization. As an art of management, innovation means that employees' personal interests are compatible with the company's interests.

Research results show that collaboration between given organizations affects innovation (especially in its management system) and will especially develop in agile organizations. Also, self-management primarily occurs in creative organizations such as scientific and artistic teams, construction offices, editorial offices, and advertising agencies. Organizations of this type are conducive to shaping creative environments that lead to the development of collaboration and the creation of networks and clusters.

Computerization in Teal Organizations

In today's changing economy, IT development is of crucial importance. On the other hand, innovation is a driver of the modern economy. The development of computer science enables the coordination of all human activities in real time, which provides information on the current state of affairs (IoT). The development of this area creates a demand for specific technological possibilities, as is the case on IoT (which calls for more and more new 5G IT systems) (Olesiński, 2020).

The modern management methods found in organizations promote self-organization, mutual trust, the formation of strong teams, and a departure from the directive management method. However, many managers still focus their efforts on finance, investments, IT systems, marketing, sales, automation, robotization, Industry 4.0, and other technical innovations and tend to forget about people—the key aspect through which a company can thrive. Therefore, attention was paid in the aforementioned studies to the adaptation of employees to situations that lead to the creation and acquisition of characteristics that indicate the type of organization (see Figure 1.2).

The results of the conducted research show that the key elements that determine the adaptation of employees to given situations are (1) the change of current work plans in a situation when the necessary resources are suddenly unavailable and (2) a sudden transition from working on one project to another.

The above-mentioned situations lead to changes in an organization and result in a different management style. However, there is not always a smooth adaptation to certain conditions (r), which affects the building of relationships, trust, or even the exchange of information.

The focus on development and computerization in an organization results from both the specifics of organizations and (above all) the

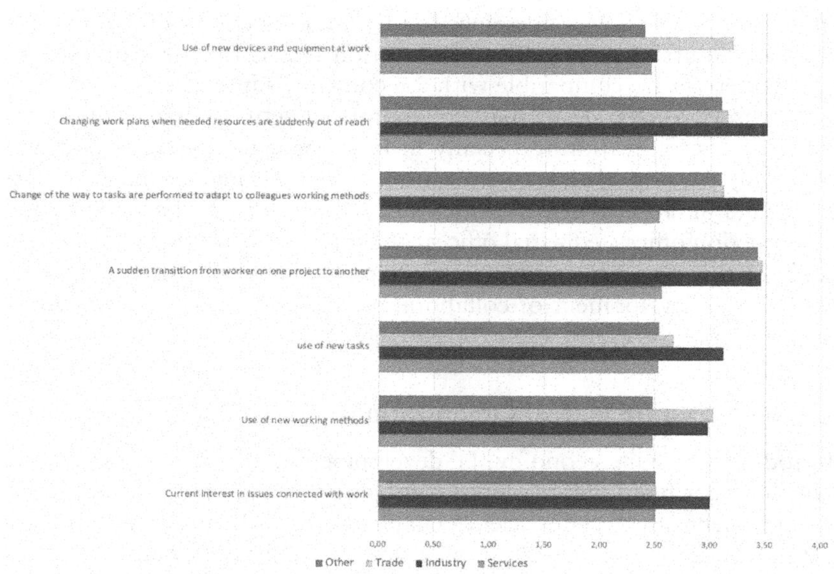

Figure 1.2 Rate at which employees of surveyed companies adapt to given situations (N=202)

Source: Rzepka, 2019

management style presented, which translates into the implementation and application of the principles found in Teal organizations.

Conclusions

The considerations presented in this chapter lead to the unequivocal but cautious acceptance of the statement that self-management relationships in Teal organizations are part of the intensification of management in the economy of automation and artificial intelligence. The role and importance of agile organizations are growing, while inter-organizational collaboration and soft factors in management in correlation with innovation and computerization only confirm and clarify this. Hence, the conviction that the five auxiliary questions and main questions are correct is justified.

According to the authors of an article in Harvard Business Review, up to 20% of organizations may have a self-managing structure (holacracy) by 2030, which confirms the earlier suggestions of F. Laloux.

The conclusion drawn from the previous considerations shows that organizing is a phenomenon that is much broader than one organization and may manifest itself in inter-organizational networks or clusters. Empirical research indicates the rapid development of network structures that is manifested by the rapid emergence of new types of consultancy, outsourcing, and other organizations as well as a significant enrichment of inter-organizational relationships through the development and differentiation of soft factors in management and increases in the complexity of the dimensions of inter-organizational relationships.

The concept of the development of an entrepreneurial agile organization is a real challenge we are facing today, both in the practice and the theory of management science. A company's development is becoming one of the strategic areas which is most often associated with the concept of entrepreneurship and where the need for the long-term research of this process is strongly emphasized. A unique combination of factors such as people, opportunities, and resources formed at a specific moment serves as a platform for organizational development. As research shows, the evolution of management toward Economy 4.0 is inevitable and even expected; we are already seeing its first symptoms.

Recommendations for Practical Application by Entrepreneurs and Scientists

The above empirical research results allow us to formulate the following recommendations.

1. It seems reasonable to constantly monitor the functioning of organizational structures and the principles on which employee teams operate. In practice, it is easier to make adjustments to such rules on an ongoing basis than to reorganize them, which does not necessarily

lead to the desired results. Generally, as young and educated employees are introduced to an employee team, they are interested in a greater freedom in the company and the freedom to introduce innovations. Hence, a flexible management position is desirable for changes in the direction of self-organization.

2. Creating the possibility of implementing process management is justified as new techniques and technologies are introduced in a functioning organization. This will create the possibility of so-called horizontal collaboration (i.e., at the level of a given hierarchy level); this will provide the opportunity to cooperate with parallel organizational structures. The development of collaboration between regular colleagues is of particular importance.
3. Shaping employee teams is a significant challenge for managers. In organizations with several or several dozen employees and several teams, employees should be introduced so as not to generate conflicts. The aforementioned results of the research on employee attitudes toward agility indicate a different degree of tolerance for changes by individuals. Therefore, the proper selection of employees requires significant experience. Creating a new team from new and young employees can be a partial solution. However, this solution can generate conflict between new and old teams.
4. The successive implementation of new IT solutions will be a significant challenge in Economy 4.0. In particular, it seems appropriate to make changes in a gradual but continuous way, which allows older workers to adapt to any changes more easily. It is desirable to promote the need for change; however, with assurances by management that the changes will be implemented in a balanced and stable manner while taking into account the adaptability to changes in older workers.

Fully automated machines with elements of artificial intelligence will pose a special challenge for management and employees. A thorough analysis of the costs of using such devices and the preparation of appropriately trained employees (preferably young and with adequate education) are necessary when implementing such solutions. In some cases, it is advisable to form separate employee teams; this will facilitate a young employee's adaptation and, for others, alleviate the stress of dealing with a new product or the need to master new skills, habits, or knowledge for any new products and processes.

References

Borowiecki, R., & Rojek, T. (2015). *Kształtowanie Relacji Partnerskich i Form współdziałania współczesnych przedsiębiorstw: Strategie, Procesy, narzędzia*. Fundacja Uniwersytetu Ekonomicznego.

Elsbach, K. D., & Kramer, R. M. (2015). *Handbook of qualitative organizational research: Innovative pathways and methods*. Routledge.

Glugiewicz, E., & Gruchman, B. (1998). The role of innovations in regional economic restructuring in eastern Europe. *High Technology Industry and Innovative Environments*, 221–232.

Grabowska, M. (2014). Współdziałanie Przedsiębiorstw w Perspektywie Sieciowej. *Zeszyty Naukowe. Organizacja i Zarządzanie/Politechnika Śląska, 76*, 51–56.

Harari, Y. N. (2019a). *21 Lekcji Na XXI Wiek* (M. Romanek, Trans.). Wydawnictwo Literackie.

Harari, Y. N. (2019b). *Homo Deus: Krótka Historia Jutra* (M. Romanek, Trans.). Wydawnictwo Literackie.

Juchniewicz, M., & Grzybowska, B. (2010). *Innowacyjność mikroprzedsiębiorstw w Polsce.* Polska Agencja Rozwoju Przedsiębiorczości.

Kidd, P. T. (1995). *Agile manufacturing: Forging new frontiers.* Addison-Wesley.

Nowacki, R. (2010). *Innowacyjność w zarządzaniu a konkurencyjność przedsiębiorstwa.* Difin.

Olesiński, Z. (2020). *Składniki Turkusowych Organizacji.* Difin.

Olesiński, Z., & Rzepka, A. (2017). *Zarządzanie Międzyorganizacyjne w Zwinnych Przedsiębiorstwach.* Texter.

Oliver, C. (1990). Determinants of interorganizational relationships: Integration and future directions. *The Academy of Management Review, 15*(2), 241–265.

Roblek, V., Meško, M., & Krapež, A. (2016). A complex view of Industry 4.0. *SAGE Open, 6*(2), 1–11.

Rzepka, A. (2018). *Relacje Międzyorganizacyjne i Kapitał Intelektualny Jako Czynniki Rozwoju Mikro i Małych Przedsiębiorstw: Studium Na Przykładzie Wybranych Przedsiębiorstw Polskich i Gruzińskich.* Difin.

Rzepka, A. (2019). Relacje Międzyorganizacyjne w Zwinnych Przedsiębiorstwach. In A. Jaki & S. Kruk (Eds.), *Zarządzanie Restrukturyzacją: Innowacyjność i Konkurencyjność w Obliczu Zmian.* TNOiK Dom Organizatora.

Schwabe, K. (2018). *Czwarta Rewolucja Przemysłowa.* Studio Emka.

Thames, L., & Schaefer, D. (2016). Software-defined cloud manufacturing for Industry 4.0. *Procedia CIRP*, 52, 12–17.

Witkowski, J. (2010). *Zarządzanie łańcuchem Dostaw: Koncepcje, Procedury, doświadczenia.* Polskie Wydawnictwo Ekonomiczne.

Zakrzewska-Bielawska, A., & Walecka, A. (2013). Organizacja w Procesach Zmian—w Drodze Do Elastyczności i Innowacyjności. In A. Adamik (Ed.), *Nauka o Organizacji. Ujęcie Dynamiczne* (pp. 294–338). Oficyna a Wolters Kluwer.

2 Challenges in Production Management in Context of Industry 4.0

Iwona Pisz and Sabina Kauf

Introduction

Since the mid-18th century, we have observed permanent changes in production methods and the increasing industrialization of the world. Technological progress causes changes in those paradigms referred to as "industrial revolutions." There has been a shift from production mechanization through mass production, electrification, and automation to digital integration. This is called Industry 4.0 (Figure 2.1); it will entail substantial changes in the implementation of production processes and in the functioning of all supply chains (Friedlmaier et al., 2018; Hackius & Petersen, 2017). Due to the potential of the reorganization of Industry 4.0, the existing methods of manufacturing products will change (Casey & Wong, 2017).

Today, we are on the verge of a new industrial revolution. In this revolution, digital networks are linked to the operational values of an intelligence factory that covers everything from the initial idea through the design, development, and production to maintenance, service, and recycling. Industry 4.0 includes the horizontal integration of data flow among partners, suppliers, and customers. The vertical integration of an organization—from development to the final product—is characteristic of this concept. The Fourth Industrial Revolution combines the virtual and real worlds; the result is a system in which all processes are fully integrated—we can observe an information system in real time. The technologies standing behind Industry 4.0 (such as artificial intelligence (AI), cyber-physical systems (CPS), the Internet of Things (IoT), digital twin, augmented reality, additive manufacturing, and cloud computing) allow decision-makers to monitor physical processes and make smart real-time decisions. The speed and pace of the changes in consumer trends will be an important driver of Industry 4.0 (Diego et al., 2019).

The 4.0 concept is becoming more and more popular nowadays (Kumar et al., 2019). However, despite the growing interest, there is still no consensus as to the new paradigm of production processes nor the

DOI: 10.4324/9781003186373-2

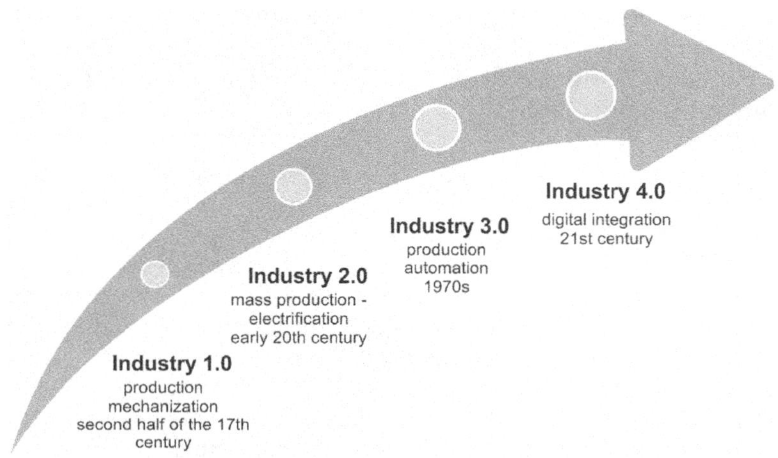

Figure 2.1 Four industrial revolutions

Source: own study

consequences resulting from the use of new technologies. In addition, many companies do not seem to be aware of the challenges that come with implementing Industry 4.0. Therefore, this chapter attempts to identify the concept in the context of the previously existing definitions in the literature as well as to identify key challenges that company managers may face (in particular, in the design and implementation of new business models). In the summary, we present our conclusions and propose our directions for future research in this area.

Research Methods

Despite the growing interest in Industry 4.0, one generally accepted definition of this concept has not been developed. In the body of works on the subject, many studies that identify the main characteristics of the Fourth Industrial Revolution (Liao et al., 2017) based on a systematic analysis of the literature can be found. Therefore, the aim of this publication is not to create another systematic analysis of previously existing works but to present the results of these studies. This allows us to identify the key characteristics of Industry 4.0, which are presented in a word cloud. This is a kind of summary of the diversity of the definitions and the basis for identifying the essence of Industry 4.0.

Next, the classic process of production management is characterized, and any changes that have occurred (or will occur) as a result of the implementation of solutions of the Fourth Industrial Revolution are identified. After this, the challenges that production management may face in the age of Industry 4.0 are presented. In the end, the potential of the research directions in the field of production management in the age of Industry 4.0 is discussed.

Industry 4.0—Concept and Goals

Since the emergence of the concept of Industry 4.0 (in 2011), it has become one of the key areas of research and implementation. The concept of Industry 4.0 is treated as a synonym of a set of new technologies initiating groundbreaking changes in the industry. It is believed that this will revolutionize the existing production methods, allow for a rapid increase in the efficiency of enterprises, and help create new business models, services, and products. The appearance of the Fourth Industry Revolution is a consequence of the dynamically changing business environment and constantly growing customer requirements. Currently, flexibility and response to changing market needs in real time are becoming increasingly important. This is why enterprises strive to adapt production processes in such a way that it is possible to produce individualized products at a time when there is a real demand for them. Industry 4.0 opens up new opportunities for creating added value for the customer and stimulating technological and process innovation, which allows for increased competitiveness. Their development will be accompanied by increasing the robotization and autoionization of production, among other things. Implementing the idea of Industry 4.0 will require cooperation and the sharing of knowledge with partners in supply chains as well as customers.

The term "Industry 4.0" was created as a part of the German government's work on identifying and analyzing upcoming changes of strategic importance for the German economy. Officially, the term was used for the first time at the Hanover Fair in 2011 as a part of a presentation on the future of the industry. In 2013, a working group of representatives of the German business industry and scientists published recommendations for the implementation of the "Strategic initiative Industry 4.0" program, which was included in the long-term project "High-Tech Strategy 2020 for Germany" and served as the foundation for the creation of the world's first Industry 4.0 platform.

The growing interest in the concept of Industry 4.0[1] has resulted in the growing number of articles and reports dedicated to it. A number of definitions are being created, but none has yet been established as the generally accepted one. What is more, the multitude of definitions makes it increasingly difficult to correctly interpret the concept and determine its essence. The Industry 4.0 definitions that were most frequently cited in the literature are listed in Table 2.1.

Table 2.1 Selected definitions of Industry 4.0

Qin et al. (2016)	Industry 4.0 encourages manufacturing efficiency by collecting data smartly, making correct decisions, and executing decisions without any doubts. By using the most advanced technologies, the procedures of collecting and interpreting data will be easier. The interoperability operating ability acts as a "connecting bridge" to provide a reliable manufacturing environment in Industry 4.0. This overall consciousness gives Industry 4.0 the most important aspect of artificial intelligent functions.
Schumacher et al. (2016)	Industry 4.0 is surrounded by a huge network of advanced technologies across the value chain. Service, Automation, Artificial Intelligence Robotics, the Internet of Things, and Additive Manufacturing are bringing in a brand new era of manufacturing processes. The boundaries between the real world and virtual reality are getting blurrier and causing a phenomenon known as cyber-physical production systems (CPPS).
Schwab (2016)	Industry 4.0 is differentiated by a few characteristics of new technologies; for example, physical, digital, and biological worlds. The improvement in technologies is bringing significant effects on industries, economies, and governments' development plans. Schwab pointed out that Industry 4.0 is one of the most important concepts in the development of global industry and the world economy.
Wang et al. (2016)	Industry 4.0 makes full use of emerging technologies and the rapid development of machines and tools to cope with global challenges in order to improve industry levels. The main concept of Industry 4.0 is to utilize advanced information technology to deploy IoT services. Production can run faster and more smoothly with minimum downtime by integrating engineering knowledge. Therefore, the product built will be of better quality, production systems are more efficient, easier to maintain, and achieve cost savings.
Mrugalska and Wyrwicka (2017)	The modern and more sophisticated machines and tools with advanced software and networked sensors can be used to plan, predict, adjust, and control the societal outcome and business models to create another phase of value chain organization, and it can be managed throughout the whole cycle of a product. Thus, Industry 4.0 is an advantage to stay competitive in any industry. To create a more dynamic flow of production, optimization of the value chain has to be autonomously controlled.
Geissbauer et al. (2016)	Industry 4.0—the Fourth Industrial Revolution—focuses on the end-to-end digitization of all physical assets and integration into digital ecosystems with value chain partners.
Pfohl et al. (2015)	Industry 4.0 is the sum of all disruptive innovations derived and implemented in a value chain to address the trends of digitalization, automatization, transparency, mobility, modularization, network collaboration, and socializing of products and processes.

(*Continued*)

Table 2.1 (Continued)

Hermann et al. (2015)	Industry 4.0 is a collective term for technologies and concepts of value chain organization. Within the modular structured Smart Factories of Industry 4.0, cyber-physical systems (CPS) monitor physical processes, create a virtual copy of the physical world, and make decentralized decisions. Over the Internet of Things (IoT), CPSs communicate and cooperate with each other and humans in real time. Via the Internet of Services (IoS), both internal and cross-organizational services are offered and utilized by participants of the value chain.

Source*:* own study based on Mrugalska and Wyrwicka, 2017; Qin et al., 2016; Schumacher et al., 2016; Schwab, 2016; Wang et al., 2016; Pfohl et al., 2015; Hermann et al., 2015; Geissbauer et al., 2016

An analysis of the definitions allows one to conclude that the majority of Industry 4.0 authors define it through the key technologies used in modern enterprises, such as cyber-physical systems (CPS), the IoT, artificial intelligence, or the Industrial Internet. Similar conclusions were also drawn by scientists from the Technical University of Dortmund, who synthesized all of the most important publications in this field (Herman et al., 2015). Their analysis shows that Industry 4.0 is most often described by its key attributes, such as smart factory, big data, the IoT, the Internet of Service, smart factory, and artificial intelligence (Figure 2.2).

As with the Fourth Industry Revolution, Industry 4.0 is a combination of technical components and the main principles of designing the processes of production so as to achieve the vertical and horizontal integration of networks of value creation (Saurabh et al., 2018). Some definitions also indicate cost and efficiency aspects. Industry 4.0 has enormous potential to reorganize not only individual enterprises but also entire supply chains. New technologies will help develop innovative products and self-organizing production processes.

The concept of Industry 4.0 means the connection of production machines in the real world with the virtual world of the Internet and information technology. People, machines, and IT systems automatically exchange information during production—within the factory and various IT systems operating in an enterprise. Industry 4.0 covers the entire value chain: from ordering and delivering components for ongoing production to shipping goods to customers and after-sales services.

Industry 4.0 provides access to every useful bit of information at any time from anywhere. This allows for the economical production of personalized products and short series (the so-called mass customization). As a result, the implementation of Industry 4.0 allows for reductions of production costs and flexible responses to customer inquiries.

Figure 2.2 Word cloud describing Industry 4.0

Source: own study based on Hermann et al. (2015)

Industry 4.0 and Changes in Production Management

The concept of Industry 4.0 is presented as an emerging structure in which manufacturing and logistics systems in the form of a cyber-physical production system (CPPS) intensively use the globally available information and communications network for an extensively automated exchange of information and in which the production and business processes are matched (Bahrin et al., 2016). The four main drivers of Industry 4.0 are the IoT, the Industrial Internet of Things (IIoT), cloud-based manufacturing, and smart manufacturing, which help to transform the manufacturing process into a fully digitized and intelligent one (Erol et al., 2016). In production management, Industry 4.0 is very responsive and adaptive to customer demands. Fulfilling individual customer needs is the main objective of Industry 4.0 (Neugebauer et al., 2016). The need for Industry 4.0 in production management is to convert regular machines to self-aware and self-learning machines to improve their overall performance and maintenance management with the surrounding interaction (Lee et al., 2014).

The impact of Industry 4.0 will be in the area of production management. The changes will be observed in research and development, designing, inventory management, service, and customer care. Industry 4.0 will transform the design, manufacture, operation, and service of products and production systems. Manufacturing will be transformed from single automated cells to fully integrated automated facilities that can communicate with one another and boost flexibility, speed, productivity, and quality. In a production system, the nine pillars of Industry 4.0 will transform isolated and optimized cell production into a fully integrated, automated, and optimized production flow. This will lead to greater efficiency and changes in traditional production relationships among suppliers, producers, and customers as well as between humans and machines. In Industry 4.0, these transformations, sensors, machines, workpieces, and IT systems will be connected along the value chain beyond a single enterprise. These connected systems (also referred to as cyber-physical systems) can interact with one another using standard Internet-based protocols and analyze data to predict failure, configure themselves, and adapt to changes. Industry 4.0 will make it possible to gather and analyze data across machines, enabling faster, more flexible, and more efficient processes to produce higher-quality goods at reduced costs. This in turn will increase manufacturing productivity, shift economics, foster industrial growth, and modify the profile of the workforce, ultimately changing the competitiveness of companies and regions (Rüßmann et al., 2015). Industry 4.0 aims at the construction of an open and smart manufacturing platform for industrial-networked information applications (Bahrin et al., 2016). Real-time data monitoring (tracking the statuses and positions of products as well as holding the instructions to control production processes) are the main needs of Industry 4.0 (Almada-Lobo, 2015).

The significant efficiency of Industry 4.0 can be obtained mainly through the consequent digital integration and intelligentization of manufacturing processes (Zhou, 2013). As we know, integration needs to take place on the horizontal axis (across all participants in the entire value chain) and on the vertical axis (across all organizational levels). Fully integrated and networked factories, machines, and products then need to act in intelligent and partly autonomous ways that require minimal manual interventions. By converting the analog data in Industry 4.0 into digital data, the information available in this productivity chain can be used by all players from any location and at any time. On this basis, production and sales processes can be optimized.

The key feature of Industry 4.0 is the smart interconnection of the products and processes of industrial production, automation, information, and communication technology (ICT) to integrated value-added chains. On one hand, more manufactured products will be smart products in Industry 4.0; they will incorporate self-management capabilities based on connectivity and computing power. On the other hand, manufacturing equipment will turn into cyber-physical production systems—software-enhanced machinery with their own computing power leveraging a wide range of embedded sensors and actuators (beyond connectivity and computing power). These systems will know their states, their capacities, and their different configuration options and will be able to take decisions autonomously—decentralized decision-making. As mass production gives way to mass customization, each product at the end of the supply chain has unique characteristics that are defined by the end customer.

The supply chains of Industry 4.0 will be highly transparent and integrated, and the physical flows will be continuously mapped on digital platforms; this will make each individual service provided by each CPPS available to accomplish the needed activities to create each tailored product. Industry 4.0 advocates that the shop floor will become a marketplace of capacity (supply) represented by the CPPS and production needs (demand) represented by the CPS. Hence, the manufacturing environment will organize itself based on a multi-agent-like system. This decentralized system with competing targets and contradicting constraints will generate a holistically optimized system, ensuring that only efficient operations will be conducted (Almada-Lobo, 2015).

Smart factories (which will be at the heart of Industry 4.0) will take information and communication technology on board for an evolution in the supply chain and production line, which will bring much higher levels of both automation and digitization. This means that machines will use self-optimization, self-configuration, and even artificial intelligence to complete complex tasks in order to deliver vastly superior cost efficiencies and better quality goods or services. Industry 4.0 ultimately aims to construct an open and smart manufacturing platform for industrial-networked information applications. It will eventually enable manufacturing firms of all sizes to gain easy and affordable access to modeling and analytical technologies that can be customized to meet their needs.

More production-related undertakings in Industry 4.0 will require increased data sharing across sites and company boundaries. Big data and analytics allow for the collection and comprehensive evaluation of data from many different sources and customers to support real-time decision-making, optimize production quality, save energy, and improve equipment service. Simulations will leverage real-time data to mirror the physical world in a virtual model, which can include machines, products, and humans. This allows operators to test and optimize the machine settings for the next-in-line product in the virtual world before the physical changeover, thereby driving down machine setup times and increasing quality. Additive manufacturing methods will also be widely used in Industry 4.0 to produce small batches of customized products that offer construction advantages such as complex lightweight designs. High-performance and decentralized additive manufacturing systems will reduce transport distances and the stock on hand (Bahrin et al., 2016).

According to product planning as a crucial decision in the implementation of Industry 4.0, process planning should also be automated with every aspect of the manufacturing process becoming automated and, therefore, interconnected to all of the different parts of the supply chain as well as the process itself. Thus, process planning morphs into a wider term—product planning.

Companies have two pathways ahead of Industry 4.0 in the area of production management (i.e., users of Industry 4.0 or providers of

Industry 4.0). Users are those who primitively try to implement CPS-based solutions in various departments, whereas providers are the ones who provide these solutions to other companies (Kagermann et al., 2013).

Challenges of Implementing Industry 4.0 in Production Management

While implementing the techniques of Industry 4.0 in production, management may face certain challenges such as the following (Luthra & Mangla, 2018; Tupa et al., 2017; Sanghavi et al., 2019):

- Upgrading existing machines;
- Capital intensiveness;
- Data-processing errors (Faheem et al., 2018);
- Worker compatibility with new technology;
- Cyber-attack—sensitivity and susceptibility of data (Lezzi et al., 2018);
- Availability of standard and benchmarked processes is very low due to the fresh concept of Industry 4.0 (Kamble et al., 2018);
- High transmissibility of data collected from a system without any quality loss (even after implementation of Industry 4.0) (Khan et al., 2017);
- Automation replacement of shortage of low-cost labor (Li, 2018);
- Environmental impact (due to automation, many energy resources used due to which non-renewable resources are depleted at a faster rate);
- New business models are required, and unskilled management skills could be a greater challenge (Moktadir et al., 2018).

Smart factories represent the connection between digital and physical production networks known also as cyber-physical systems. In particular, the integration of computing, wireless, and Internet technologies makes this connection possible. The IoT is the most critical component for connecting devices without wired connections. For this reason, some of the greatest challenges for manufacturing companies are to increase the level of digitalization, adapt the production lines to new technologies, and define the role of humans within the new processes.

The key challenge for Industry 4.0 is the transformation of classic machines and devices into self-aware and self-learning machines. Undistorted by interactions with the environment, they will allow us to improve the efficiency of production processes and management (Lee et al., 2014). The challenge of Industry 4.0 is also to create an open and intelligent production platform that allows for real-time data monitoring, the tracking of product status and position, and the storing of control instructions of the production process. All of this allows for the adaptation to dynamically changing market requirements (Wang et al., 2016).

However, the implementation of Industry 4.0 is associated with the need to solve many problems; the key ones can be considered as follows:

1. Investment outlays. The implementation of new technologies requires capital commitment and huge financial outlays that many manufacturing companies cannot afford.
2. Intelligent production decision-making systems. These should be self-organizing and self-learning (Wang et al., 2016). This requires the use of cognitive intelligence capable of effective thinking, especially in uncertain and unpredictable situations.
3. Ultra-fast 5G wireless technology. Without this, it will be impossible to fully exploit the potential of the IoT. Current data transmission networks often have insufficient bandwidth and do not allow for the transfer of large data sets from production processes. Thus, a huge challenge is to ensure the high quality and integrity of the data from the production system (compatible with other cells in the creation of value chains).
4. The problems specific to production process analytical tools. These are important because the amount of information obtained from machines and production systems will increase exponentially. This results in the need to use intelligent applications that allow for inference and the presentation of relevant analyses. The latter should be different, depending on the organizational level of the enterprise (Wang et al., 2016).
5. System modeling and analysis. Here, the key is to reduce the dynamic equations and formulate a suitable model for controlling machines and self-organizing devices in the production system.
6. Cyber Security. With the increasing number of connections and the use of still-standard communication protocols, the demand for critical data protection from smart factories and system data against cyber threats is increasing.
7. Bearing in mind the above problems, several recommendations and suggestions can be made for industry that management could include in their digitization strategies: (1) the further deepening of practical knowledge in the field of digital production technology and the implementation of appropriate digital production strategies; (2) integrating a digital transformation not only with a company's operational departments but also (and perhaps above all) with a company's global strategy. This approach will minimize the risk of incorrect current decisions and facilitate contact and communication with the employees; (3) guaranteeing data security, not only on the scale of an individual enterprise but throughout the entire value-creation chain; (4) integrating early with partners, which will

allow for the establishment of appropriate technical standards; (5) stimulating a culture of cooperation characterized by flexibility and openness; (6) implementing a training system (including e-learning) that raises the competences and qualifications of the employees (due to which, it will be possible to adapt the skills of the human capital to the requirements of Industry 4.0); (7) constructing flat and decentralized organizational structures that allow for increased operational efficiency; (8) establishing interdisciplinary project teams consisting of engineers, programmers, marketing, logistics specialists, etc.; (9) developing new models of working times and methods of remunerating employees, which will stimulate creative thinking and co-create value (for clients and the enterprise); and (10) introducing such information systems and data exchanges between participants of the value-creation process that will be based on openness with and trust in all partners.

In summary, the key is to develop new business models that will replace the existing ones and be able to spread groundbreaking innovations. The models should be based on innovative approaches to values and promote the intensification of customer relationships.

Conclusions

The considerations contained in this chapter focused on the Fourth Industrial Revolution, (referred to in the German-language literature as Industrie 4.0, and the Anglo-Saxon—smart industry or intelligent manufacturing). This technology allows for intelligent, efficient, individualized, and customized production at a reasonable cost level. Automation, robotization, digitization, and artificial intelligence will allow for the fast transmission and analysis of large data sets (big data). The relationships among machines allow them to communicate and learn from each other. This chapter indicates the changes that we can observe in production processes under the influence of Industry 4.0 as well as the challenges faced by modern enterprises implementing the technologies of the Fourth Industrial Revolution.

The presented considerations allow one to conclude that the 4.0 environment is more and more often the subject of scientific research and practical implementations. However, there is still a need to identify areas of application of the new technologies. It mainly gathers information from various links in the value chains so that it is possible to optimize production processes and effectively use Industry 4.0 instruments such as cyber-physical systems, the IoT, and cloud computing. It would seem that Industry 4.0 is reflected in the robotization and autonomization of manufacturing (in which man becomes superfluous). In practice, however, this is not the case. Due to Industry 4.0, smart factories are offering better jobs; and due to new technologies and solutions, employees can receive greater support than ever before. What is more, there is even the term "Industry 5.0," which means the Fifth Industrial Revolution. This intends to bring synergistic cooperation between people and machines and allow for the further personalization of products. Its idea is a systematic transformation that promotes a civil society and a return to the human touch of manufacturing, that is, increasing the degree of cooperation between people and intelligent production systems. Such a relationship helps ensure a combination of the best features of two worlds: (1) the speed and accuracy guaranteed by automation with (2) the cognitive skills and critical thinking of people.

Industry 4.0 is focusing on new technologies; the next stage in the development of smart factories is to focus on using the intellectual potential of production workers.

Industry 4.0 is part of the global megatrends of digitalization, whose significance is increasing in all areas of life as well as the economy. Industry 4.0 brings some challenges regarding data security, finding the necessary capital, developing a strategy for implementing it, and finding qualified employees (Schröder, 2016). This concept creates some opportunities for manufacturing companies (including small and medium-sized companies). Applying Industry 4.0 in companies can increase their flexibility, productivity, and competitiveness (Kagermann et al., 2013). In Germany's advanced manufacturing landscape, for example, Industry 4.0 can drive productivity gains of 5–8% on the total manufacturing costs over 10 years, totaling €90 to 150 billion. The connectivity and interaction among parts, machines, and humans will make production systems as much as 30% faster and 25% more efficient, elevating mass customization to new levels (Rüßmann et al., 2015).

Industry 4.0 poses a number of challenges for the Polish industry. In the report "Digital Poland" developed by McKinsey (Broniatowski, 2016), a much lower degree of digitization is indicated in Poland as compared to the United States or Western Europe. Poland's "digitization index" is 34% lower than that of Western Europe. In addition, McKinsey points out that the "digitization gaps" in operational sectors such as "advanced industrial production" and "simple industrial production" are at 45 and 78%, respectively, as compared to Western Europe. Research on the level of automation of Polish production collections shows that only 15% of its factories are fully automated and as many as 76% indicate partial automation (Hajkuś, 2015). Moreover, the density of robotization in Poland is almost 15 times lower than in Germany and four times lower than the average in the overall table according to the International Federation of Robotics (IFR report, 2015). In addition, a small proportion of the factories still use IT systems for production management and control (FEM). Due to this research, it is evident that the challenges of the Fourth Industrial Revolution are still valid for managers of Polish factories.

Note

1. Although the concept of Industry 4.0 comes from the German language area, its equivalents in Anglo-Saxon literature are terms such as smart industry and smart manufacturing, advanced manufacturing, intelligent manufacturing, and connected factories.

References

Almada-Lobo, F. (2015). The industry 4.0 revolution and the future of Manufacturing Execution Systems (MES). *Journal of Innovation Management JIM*, *3*(4), 16–21.

Bahrin, M. A. K., Othman, M. F., Azli, N. H. N., & Talib, M. F. (2016). Industry 4.0: A review on industrial automation and robotic. *Jurnal Teknologi*, *78*(6–13), 137–143.

Broniatowski, M. (2016). *Cyfrowa Polska. Szansa na skok technologiczny do globalnej pierwszej ligi gospodarczej*. McKinsey & Company, Forbes (In Polish).

Casey, M., & Wong, P. (2017). Global supply chains are about to get better, thanks to blockchain. *Harvard Business Review*.

Diego, G. P., Pasquale, D., & Uday, K. (2019). The Industry 4.0 architecture and cyber-physical systems. In *Handbook of Industry 4.0 and SMART systems*. CRC Press.

Erol, S., Jäger, A., Hold, P., Ott, K., & Sihn, W. (2016). Tangible industry 4.0: A scenario-based approach to learning for the future of production. 6th CLF — 6th CIRP conference on learning factories. *Procedia CIRP*, *54*, 13–18.

Faheem, M., Shah, S. B. H., Butt, R. A., Raza, B., Anwar, M., Ashraf, M. W., Ngadi, M. A., & Gungor, V. C. (2018). Smart grid communication and information technologies in the perspective of Industry 4.0: Opportunities and challenges. *Computer Science Review*, *30*, 1–30.

Friedlmaier, M., Tumasjan, A., & Welpe, I. M. (2018). *Disrupting industries with blockchain: The industry, venture capital funding, and regional distribution of blockchain ventures*. Proceedings of the 51st Hawaii International Conference on System Sciences, Big Island, HI, USA, pp. 3517–3526.

Geissbauer, R., Vedso, J., & Schrauf, S. (2016). *Industry 4.0: Building the digital enterprise*. www.pwc.com/gx/en/industries/industries-4.0/landing-page/industry-4.0-building-your-digital-enterprise-april-2016.pdf.

Hackius, N., & Petersen, M. (2017). *Blockchain in logistics and supply chain: Trick or treat?* Proceedings of the Hamburg International Conference of Logistics (HICL), Hamburg, Germany, October 12–14, pp. 3–18.

Hajkuś, J. (2015). *W jakie technologie inwestują firmy produkcyjne?* Raport ASTOR (In Polish).

Hermann, M., Pentek, T., & Otto, B. (2015). *Design principles for industrie 4.0 scenarios: A literature review*. www.researchgate.net/publication/307864150_DesignPrinciples_forIndustrie40_ScenariosA_LiteratureReview

IFR International Federation of Robotics. (2015). *World robotics 2015 (report)*. IFR.

Kagermann, H., Wahlster, W., & Helbig, J. (2013). *Recommendations for implementing the strategic initiative INDUSTRIE 4.0*. Final Report of the Industrie 4.0 Working Group. Acatech.

Kamble, S. S., Gunasekaran, A., & Sharma, R. (2018). Analysis of the driving and dependence power of barriers to adopt industry 4.0 in Indian manufacturing industry. *Computers in Industry*, *101*, 107–119.

Khan, M., Wu, X., Xu, X., & Dou, W. (2017). *Big data challenges and opportunities in the hype of industry 4.0*. IEEE International Conference on Communications (ICC), Porte Maillot, Paris, France, pp. 1–6.

Kumar, K., Zindani, D., & Davim, J. P. (2019). *Industry 4.0, developments towards the Fourth Industrial Revolution*. SpringerBriefs in Applied Sciences and Technology, Singapore.

Lee, J., Kao, H. A., & Yang, S. (2014). Service innovation and smart analytics for industry 4.0 and big data environment, product services systems and value creation. Proceedings of the 6th CIRP conference on industrial product-service systems. *Procedia CIRP*, *16*, 3–8.

Lezzi, M., Lazoi, M., & Corallo, A. (2018). Cybersecurity for industry 4.0 in the current literature: A reference framework. *Computers in Industry*, *103*, 97–110.

Li, L. (2018). China's manufacturing locus in 2025: With a comparison of "Made-in-China 2025" and "industry 4.0". *Technological Forecasting & Social Change*, *135*(C), 66–74.

Liao, Y., Deschamps, F., Loures, E., & Ramos, L. F. (2017). Past, present and future of industry 4.0—a systematic literature review and research agenda proposal. *International Journal of Production Research*, *55*(12), 3609–3629.

Luthra, S., & Mangla, S. K. (2018). Evaluating challenges to industry 4.0 initiatives for supply chain sustainability in emerging economies. *Process Safety and Environmental Protection*, *117*, 168–179.

Moktadir, M. A., Ali, S. M., Kusi-Sarpong, S., & Shaikh, M. A. A. (2018). Assessing challenges for implementing industry 4.0: Implications for process safety and environmental protection. *Process Safety and Environmental Protection*, *117*, 730–741.

Mrugalska, B., & Wyrwicka, M. K. (2017). Towards lean production in industry 4.0. *Procedia Engineering*, *182*, 466–473.

Neugebauer, R., Hippmann, S., Leis, M., & Landherr, M. (2016). *Industrie 4.0 — From the perspective of applied research*. 49th CIRP Conference on Manufacturing Systems (CIRP-CMS 2016), Stuttgart, Germany, pp. 2–7.

Pfohl, H.-Ch., Yahsi, B., & Kurnas, T. (2015). The impact of industry 4.0 on the supply chain. *Proceedings of the Hamburg Inter Innovations and Strategies for Logistics*, *20*.

Qin, J., Liu, Y., & Grosvenor, R. A. (2016). Categorical framework of manufacturing for industry 4.0 and beyond. *Procedia CIRP*, *52*, 173–178.

Rüßmann, M., Lorenz, M., Gerbert, P., Waldner, M., Justus, J., Engel, P., & Harnisch, M. (2015). *Industry 4.0: The future of productivity and growth in manufacturing industries*. The Boston Consulting Group.

Sanghavi, D., Parikh, S., & Raj, S. A. (2019). Industry 4.0: Tools and implementation. *Management and Production Engineering Review*, *10*(3), 3–13.

Saurabh, V., Prashant, A., Santosh, B. (2018). *Industry 4.0—A glimpse*. 2nd International conference on materials manufacturing and design engineering, Beijing, China. *Procedia Manufacturing*, *20*, 233–238.

Schröder, C. (2016). *Herausforderungen von Industrie 4.0 für den Mittelstand*. Friedrich-Ebert-Stiftung.

Schumacher, A., Erol, S., & Sihn, W. (2016). A maturity model for assessing industry 4.0 readiness and maturity of manufacturing enterprises. *Procedia CIRP*, *52*, 161–166.

Schwab, K. (2016). *The fourth industrial revolution, what it means and how to respond*. www.weforum.org/agenda/2016/01/the-fourth-industrial-revolution-what-it-means-and-how-tores pond.

Tupa, J., Simota, J., & Steiner, F. (2017). Aspects of risk management implementation for industry 4.0. *Procedia* Manufacturing, *11*, 1223–1230.

Wang, S., Wan, J., Li, D., & Zhang, C. (2016). Implementing smart factory of industrie 4.0 an outlook. *International Journal of Distributed Sensor Networks*, *6*(2).

Zhou, J. (2013). Digitalization and intelligentization of manufacturing industry. *Advances in Manufacturing*, *1*, 1–7.

3 Challenges of Business Model Concept of Small- and Medium-Sized Enterprise Cooperation

Sebastian Saniuk and Sandra Grabowska

Introduction

The widespread automation and digitization of industry caused by the development and use of new technologies (especially the IoT, Cloud Computing, and Big Data) create significant opportunities for product innovation and value creation for companies operating in Industry 4.0 (Fatorachian & Kazemi, 2018). The Industry 4.0 concept means a close connection of geographically dispersed intelligent physical objects with the ICT network in the entire product life cycle (PLC) process which consist of the following:

- The creation of the concept of a new product;
- Virtual documentation;
- 3d printing models;
- Laboratory and industrial research;
- Preparation of virtual production documentation;
- Simulation of product manufacturing in a virtual production environment;
- Checking its correctness;
- Making decisions about production in a cyber-physical system;
- Organization of production and logistics supported by computer-aided systems;
- Preparation of production and assembly documentation;
- Organization of warehouse, transport, and sales;
- Control of the correctness of operation;
- Compliance with deadlines for inspections, repairs, and renovations;
- Indicating the place and contractor of these activities;
- Controlling recycling, often with the active participation of the customer affecting the design, production, and distribution process and product.

This means that it is possible to meet consumer expectations due to the dynamic adaptation of the autonomous modules of the entire production

DOI: 10.4324/9781003186373-3

preparation process, production, and delivery of the product to the consumer (Burmeister et al., 2015). The contemporary nature of production is shaped by the change of the paradigm from mass production to production at the customer's request. The range of customer expectations for products is so wide and diverse that only creating cooperation for the needs of a specific project will allow for building a competitive advantage of a company (especially one from the SME sector). A company's participation in the network is particularly attractive for small- and medium-sized enterprises, which in this way can overcome the main competitive advantage of large enterprises in terms of access to all kinds of resources (capital, competencies, know-how, etc.). Networking also has a positive effect on the learning process by gaining experience, know-how, and knowledge transfer based on mutual relationships between cooperating enterprises.

The study and findings presented by Andreas Schroeder et al. contribute to the development of the business network perspective in the context of Industry 4.0. In the study, the authors illustrate the many levels of interactions and dependencies that characterize enterprises in collaboration (paying particular attention to the existing resource theory) by contributing to an understanding of the specific resource aspects of product-use data and its implications. In addition, it emphasizes the importance of the growing capacity of the network and its application in the context of cooperation in Industry 4.0 (Bulger et al., 2014). Strong competition and growing customer expectations in the modern market mean that product customization occurs along with increases in production efficiency (i.e., with active customer participation). At the same time, the price of a product should be similar to the price of products manufactured in series and mass production. Customization can take different forms depending on the degree of customer influence on the new product. The highest degree of customer interference occurs in the case of pure customization, in which the customer participates in the entire production process beginning from the product-design phase and manufacturing to the distribution of the product. This organization requires the development of an appropriate interface for the client who would allow him to participate in the design and then the manufacture of the ordered product (Bauernhansl et al., 2014). An easier way to organize product customization is tailored customization. In this case, the customer is involved in the product-manufacturing phase and has an impact on the shape and dimensions of the standard product components. The customer's more limited impact on the manufacturing process can be seen in the customized standardization strategy, where the customer has an impact on assembly or distribution by choosing the available configuration options for a standard product (e.g., the configuration of a car during purchase). In all cases of customization, an important goal is to provide an individualized product at a price close to the price of the product offered in mass production (Suomala, 2002). This means the need to significantly increase production efficiency and reduce production costs. A significant increase in production efficiency becomes possible through the use of the material, production, and employee resources of cooperating network partners with unused production capacity (Grabowska et al., 2020).

In addition, developing open applications is required that enable the network connection of enterprises involved in the product-production process, delivery to the customer, and users of this product and to ensure the security of the collected and processed data. An important element of the development of the Industry 4.0 concept is also the legislative process, which requires the development of a number of new legal provisions at the EU level (at least), enabling the network exchange of production and service data to cover the entire value chain (Emiliani, 2003).

An important challenge is also to ensure the security of the collected data in distributed collections using big data and cloud computing technologies. Transferring a huge amount of information between network partners will be carried out in real time, which in turn will require additional investment in cybersecurity (especially for end-users). When designing the organization of cyber-physical flows, security-by-design, and privacy-by-design approaches should be introduced instead of the currently dominant one—involving the fastest and cheapest implementation of a given project without taking cybersecurity into account (Hustinx, 2010; Santos et al., 2017).

Modern enterprises are facing the new challenge of building business models based on cooperation within cyber-physical systems. However, this requires a change in the way they operate in terms of market functioning as well as a solution to a number of problems related to the creation and management of enterprise networks (hence, the need to conduct research related to the construction of new business models dedicated to the SME sector in the era of implementing the concept of Industry 4.0). Industry 4.0 is a big challenge for enterprises and, at the same time, huge future developmental prospects. Companies must completely change their strategies, create new business models, and overcome many operational and technological barriers (Casadesus-Masanell & Ricart, 2010).

In this chapter, the Business Model 4.0 concept of small- and medium-sized enterprises cooperating within an e-business platform that supports the creation of temporary production networks that guarantee personalized product execution in the Industry 4.0 environment is presented.

Research Methods

The research methodology is based on two stages. The first stage included our initial research, a literature review, and an analysis. The second stage is a direct study of multiple case studies. The following research techniques were used in these studies: the competent judges' method, the categorized interviews with managers and entrepreneurs method, a document analysis, and the comparative method. The results of multiple case studies helped us answer the following questions:

1. Which components should be utilized in Business Model 4.0?
2. How do you define Business Model 4.0?
3. Which Industry 4.0 technologies will affect the business model architecture?
4. How do you organize the process of forming/planning cybernetworks to carry out personalized production orders?
5. Which key performance indicators (KPI) should be used to assess Business Model 4.0 and assess the functioning of the network (CIN)?

In the adopted research methodology, the competent judges method was used to select enterprises for conducting multiple-case studies. The competent judges were three academicians dealing with the subject of business models, Industry 4.0, and the SME sector, two strategic clients of industrial enterprises, and seven entrepreneurs from the SME sector. Multiple case studies were based on a selected panel of five manufacturing companies from the SME sector. This is due to the assumption adopted in social sciences that, for some problems, it is more appropriate to use in-depth analyses of a smaller number of cases than the superficial studies of a large number.

Results

Architecture of Business Model 4.0

In a turbulent and changing environment, businesses are forced to look for the most effective methods of monitoring and detecting changes in the environment to undertake effective adaptation activities resulting from the Industry 4.0 concept. Globalization provides an opportunity to build a competitive business model that is not dependent on the size of a company, country of origin, or advantage in the past. Only business models based on establishing cooperation will be able to compete with large-market players (Drucker, 2004). The increase in production efficiency due to automation as well as effective data collection and analysis methods combined with an effective quick decision-making process means changes in the way a business is run; new business models with very high capitalization are created (Hall et al., 2017).

The idea of a production network called the cyber industry network (CIN) means the manufacturing of joint production orders using fully automated processes of individual network partners in which communication takes place via the Internet and the necessary data is stored in the cloud (cloud computing). This enables the constant access of all participants of the network to selected necessary information from anywhere in the world. Thus, the chance for development arises in creating a partnership consisting of the combination of specialized competencies and the ability to change in order to better meet customer expectations and enable the effective acquisition of a competitive advantage in the market (Saniuk et al., 2019).

This form of cooperation provides access to a variety of resources; for example, machines giving a full range of production possibilities without having to invest in their purchase. The condition for their creation is the development of a model of cooperation between enterprises creating cyber-physical systems in the future. CPSs should ensure data collection, processing, and impact on physical processes occurring throughout the entire production network due to unlimited connection networks of intelligent mechatronic resources (machines, devices, robots, means of transport, etc.) communicating with each other with negligible participation of a man who only performs the functions of supervision and/or coordination. The concept of a business model based on cooperation is presented in Figure 3.1.

The development of such a model is closely related to the need to propose a set of conditions for the functioning of an enterprise in the production network as well as to develop ways of establishing temporary networks oriented toward a joint venture. In particular, this means selecting partners for the network, planning the load of geographically dispersed resources (scheduling), controlling production, or defining the principles of the financial settlement of partners providing resources for the production. The accession of the company to such a network model operating under the concept of Industry 4.0, first of all, requires an initial assessment of its technological potential, know-how, employee competencies, and ability to sharing resources. Therefore, enterprises must ensure the following in their organizations:

- Potential capabilities to reduce the technological gap, including ensuring the so-called technological readiness;
- Adequate social and technological potential;
- Internal ability to absorb and adapt to innovation.

Each enterprise wishing to participate in building a network should decide in which area it will specialize in the needs of network cooperation. This allows one to focus one's attention on only those resources that guarantee a high level of utilization and ensure the competitiveness of the enterprise as well as reduce the very high costs of building a specialized infrastructure.

The idea of creating production networks (CIN) provides for the use of an e-business platform supported by a network broker. This solution is dedicated in particular to a group of small- and medium-sized enterprises,

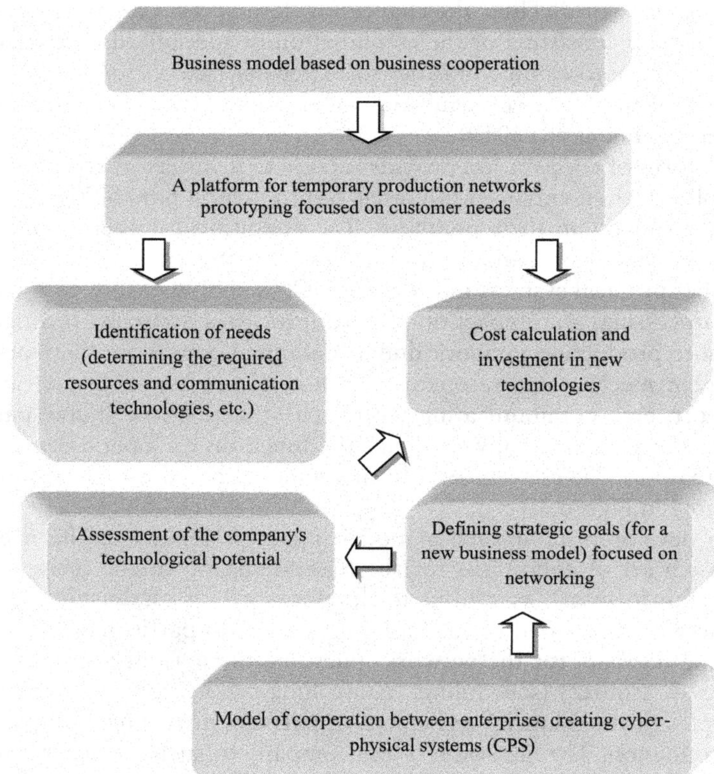

Figure 3.1 Concept of a business model based on cooperation
Sources: Authors' elaboration.

which will make available within the platform of intelligent resources communicating directly with the platform and with other resources in the network of enterprises appointed each time for the needs of the personalized product being manufactured. The proposed approach to creating a CIN is based on the exchange of information between enterprises, brokers, and clients. The broker will be responsible for supervising the collection of data on the production order and offers collected from enterprises in real time. The data from the enterprises will be automatically transferred from machines and devices using the IoT, cloud computing, and big data technologies to the e-business platform system supervised by a network broker. Based on the information collected in the system, a set of network variants guaranteeing the timely order execution and a schedule showing a load of individual company resources in the selected variant for the implementation of a new order will be generated (Grabowska et al., 2020).

The fundamental transformation of business is just taking place. Industry 4.0 will absolutely force enterprises to change their models of operation. One of the most important models of business management that enables the flexible, effective, and competitive functioning of a company on the market is a process-based model.

In order to create a business model of a network of manufacturing enterprises operating in the era of Industry 4.0, it is necessary to start with some assumptions (Prahalad & Krishnan, 2010):

- The value is determined by the experience created at a given time in cooperation with a single consumer;
- The company is not vertically integrated;
- The company is not able to satisfy the consumer's expectations at a given time due to its size and activities;
- Attention is focused on access to resources, not on the ownership of resources;
- Resources are derived from various suppliers, and access to them is global;
- It is crucial that the supply of products, services, and competencies is multi-institutional.

The observed increase in competitiveness and globalization processes as well as any ongoing mergers and acquisitions affect the search for new methods, techniques, and management instruments. The ongoing development of a knowledge-based economy expressed in the intensive transfer and diffusion of innovation has a significant impact on changes in business models and business processes. New forms of competitiveness and cooperation emerge. New models applying a wide range of different types of innovations are used—these are business models based on the principles of a "new era of innovation" (Prahalad & Krishnan, 2010).

Using the concept of the "new era of innovation" for the research needs of this work, it was assumed that the business model is a configuration of business processes that combine and develop resources that are shaped in the form of the social and technical architecture of the enterprise.

The need to conduct research in the field of the business models of enterprise networks can also be proven by the numerous benefits of business cooperation indicated in the literature. Entering enterprises into various cooperative arrangements (including the organization of permanent or temporary networks) is particularly attractive for small- and medium-sized enterprises; in this way, they can overcome the main competitive advantage of large companies, especially in terms of access to all types of resources (capital, competencies, know-how, etc.) (Afuah, 2004; Afuah & Tucci, 2001; Emiliani et al., 2003).

According to the new era of innovation, the business model is based on the configuration of the social and technological architectures of the interconnected business processes and perfectly meets the requirements of Industry 4.0 (such as product customization, the need to build cooperation networks, digitization, robotization, and Logistics 4.0).

In this model, the role of business processes is clearly emphasized. In practice, the elements of such a business model are as follows:

- Social architecture (knowledge resources, management systems, competencies, employee development, and motivation);
- Technological architecture (it and telecommunications devices, computers, ICT systems, robots, etc.);
- Business processes that combine these databases (essentially infrastructural) and at the same time derive the necessary resources from them for the implementation of the appropriate products that create value for the client.

Business Model 4.0 will be understood as the configuration of business processes combining and developing resources that are shaped in the form of the social and technical architecture of the company, built on flexible processes, and based on a virtual cooperation network that can meet the demand for personalized products.

Business Model 4.0—Canvas

A creative approach for creating innovative business models is Business Model Canvas, which consists of nine main elements: customer segments, the value offered to customers, sales and service channels, the nature of customer relationships, key processes, resources, partnerships, income, and cost structure.

Business Model Canvas (BMC) provides a simple yet comprehensive format for mapping and refining the key factors of a business. The relationship between entrepreneur and canvas is similar to that between an architect and a blueprint. Unlike a traditional business plan, it is flexible—easily adapted to changing business ideas and new information.

BMC can also be referred to as a business model template where each of the nine blocks are elements representing various aspects of a company's operation. The elements are filled with content and combined to describe the product or service that a company provides to its customers.

How does one use this model (bearing in mind the potential for radical changes that Industry 4.0 brings along the entire value chain in the manufacturing sector)? Expecting changes in the balance of power on the market, it should be borne in mind that one of the pillars of market change represents new cooperation opportunities based on digital technologies as well as the belief that cooperation in the spirit of win-win at the "end of the day" will be profitable for all participants of the transaction. The following Business Model 4.0 pillars have been identified using the Business Model Canvas:

- Key Partners: Partners working in virtual network creating agile teams to implement a specific project, customer as a partner, participant in the product-design process;
- Key Actions: Automated production in line with customer expectations, production as a service, production as a product;
- Value proposition: Product is personalized according to individual customer's order at a price of product from mass production, IoT platforms;
- Customer relations: Personal relationship with the client, relationship with use of latest technology, digital platforms, partnership throughout the entire product life cycle;
- Customer segmentation: Mass personalization, segment market, diversified market;
- Key Resources: Elimination of unused production potential, knowledge about customer preferences;
- Channels: Retail network;
- Cost structure: Management production;
- Revenue structure: Sale of products/services; fee charged for using the product; licensing.

The greatest advantage of the Business Model 4.0 Canvas template is undoubtedly its simplicity and intuitiveness. It is easy to present and very understandable. The versatility of the application, high flexibility of the discussed approach, and placement of the customer template in the center of the template are worth emphasizing; this allows an entrepreneur to prioritize their business operations.

Key Performance Indicators for Business Model 4.0

Translating the basic strategic assumptions of an organization into measurable goals is a tedious and complicated process. It is very important that the entire organization goes in one direction in order to create a profitable business model. This requires building a clear and unambiguous way of communicating goals as well as the degrees of their implementation.

Key performance indicators (KPIs) are tools that help one to manage almost any area of their business. KPIs are elements of a business plan that express a goal (Brown, 1996). These are measurable result-based statements that will be used to measure whether an enterprise is on track to achieve its goals. Good plans use five to seven KPIs to manage and track progress. The anatomy of a structured KPI includes the following (Parmenter, 2015):

- A measure—each KPI must have a measure. The best KPIs have more expressive measures;
- A target—each KPI must have a target that matches one's measure as well as the time period of one's goal. This is generally a numeric value that one seeks to achieve;
- A data source—each KPI must have a clearly defined data source so there is no gray area in how each is being measured and tracked;
- Reporting frequency—different KPIs may have different reporting needs (a good rule to follow is to report on them at least monthly).

The use of KPIs provides knowledge that allows one to monitor the levels of the implementations of the managerial assumptions on an ongoing basis—this, in turn, enables one to make quick decisions, prioritize activities, and improve strategies. These indicators allow for reducing comprehensive information about the company and its results to a small amount of key data, giving understandable results on the basis of which to draw conclusions and modify the method of operation.

In the words of Peter Drucker: "if something can be measured, then it can also be managed." However, KPIs do not mean measuring every piece of information. These are key indicators—those whose measurement provides reliable results that reflect the state of a company and the degree of the implementation of its plans (Grycuk, 2010).

Therefore, it is important to select the right indicators, as not everything that can be measured is equally important for expanding knowledge and drawing the correct conclusions. From the whole range of available options, one or several indicators should be selected that best demonstrate the level of the implementation of the strategic goals.

KPIs designated by different companies or organizations will differ due to such conditions as the industry, the company size, or its specific features. To achieve meaningful results, it is necessary to properly define the indicator in relation to the company and its specific needs and to set target values to be achieved. The indicator must relate to information that is crucial and assume an achievable goal. You also need to define the period of time during which the indicator should reach the desired value (Li et al., 2016).

Measuring results using KPIs aims to expand knowledge and improve company results. KPIs help managers define and achieve operational and strategic goals. Table 3.1 shows examples of KPIs for Business Model 4.0.

The KPIs presented in the table are quite diverse. The developed indicators are examples and should be treated as demonstrative. In order to apply them in a particular enterprise, they should be adapted to their specificity, size, and development dynamics and choose the most appropriate Indicators that correlate with the strategic goal of the enterprise. The proposed measures, sources of information, or measurement frequency may be different in individual companies. However, it should be remembered that, in order to be helpful in achieving their success, they must be reliably accounted for, reported, and controlled.

Table 3.1 KPI for Business Model 4.0

KPI/Goal	*Measure*	*Source of information*	*Measurement frequency*
Critical success factors			
Shift management	Number of changes initiated	All company departments	Once per month
	Number of implemented changes	All company departments	Every 3 months
Competency	Number of training sessions	HR department Training Department	Every 6 months
Cooperation within network	Number of networks established for the implementation of specific task/project/product	Procurement Department, Logistics, Production, Sales, Service	Once per month
	Number of orders carried out in network	Procurement Department, Logistics, Production, Sales, Service	Once per month
Automation and robotization of the production system			
Autonomous robots	Number of autonomous robots	Production Department, Logistics	Every 6 months
Simulations			
Computer programs for product modeling	Number of programs	IT Department Production Department	Every 6 months
Process simulation programs	Number of programs	IT Department Production Department	Every 6 months
Software integration			
Data freely transferred between systems	Amount of data sent	IT Department	Once per month
	Number of systems between which data is transferred	IT Department	Once per month

(*Continued*)

Table 3.1 (Continued)

KPI/Goal	*Measure*	*Source of information*	*Measurement frequency*
Programs/ systems used to communicate with the supplier	Number of programs/ systems	IT Department	Once per month
Use of electronic Kanban Cards	Number of electronic Kanban Cards used	Production Department	Once per month
Programs/systems/ tools used to communicate with client	Number of programs/ systems/ tools used to communicate with client	IT Department Production Department Sales Department Logistics Department Marketing Department	Every 6 months
Industrial Internet of Things, Internet of Services			
Application of sensors collecting information	Number of sensors installed	Production Department	Once per month
	Number of product types with sensors	Production Department	Once per month
	Number of production cells where sensors are mounted	Production Department	Once per month
Cybersecurity			
Data security level	Number of cyberattacks	IT Department	Once per month
	Number and frequency of backups performed	IT Department	Once per month
Cloud			
Cloud-based software	Number of cloud-based applications	IT Department	Once per month
	Amount of data stored in the cloud	IT Department	Once per month
Use of incremental technologies			
Product prototypes	Number of prototypes of new products made	Design Department (it can be an organizational unit in the production department)	Once per month

KPI/Goal	*Measure*	*Source of information*	*Measurement frequency*
Manufacture of products	Number of manufactured products	Production Department	Once per month
Augmented reality (e.g., glasses displaying instructions superimposed on the observed object)			
Operator support	Number of production stations equipped with glasses	Production Department	Every 3 months
Searching for materials in the warehouse	Number of production stations equipped with glasses	Production Department Warehouse	Every 3 months
Equipment repair	Number of devices equipped with glasses	Production Department	Every 3 months
	Number of failures removed using glasses	Service Department	Once per month
Virtual training	Number of training sessions conducted in virtual format	HR Department	Once per month
	Number of employees trained	HR Department	Once per month
Large data sets			
Data on product quality	Number of products manufactured	Production Department	Once per month
	Number of defective products	Production Department Quality Department	Once per month
Machine data	Number of Failures	Production Department Service	Once per month
	Machine downtime due to breakdowns	Production Department Service	Once per month
	Machine downtime due to need to change it	Production Department Service	Once per month
	Machine downtime caused by its maintenance	Production Department Service	Once per month
	Machine downtime due to its use	Production Department	Once per month

(*Continued*)

Table 3.1 (Continued)

KPI/Goal	*Measure*	*Source of information*	*Measurement frequency*
Use of collected data in real time	Amount of data used and its purpose	All departments	Once per month
Autonomous systems based on artificial intelligence	Number of such systems	All departments	Every 6 months
Digital process twin that allows for ongoing comparison of the process with the reference model			
Changes made to process	Number of changes introduced in the process Number of processes changed	All departments	Every 6 months
Customization			
Product's pure customization Product's tailored customization Product's customized standardization Product's pure standardization	Number of products offered in each option Number of products produced in each option	Marketing Department Sales Department Production Department	Once per month
Artificial Intelligence (machine learning)			
Autonomous process implemented by artificial intelligence The process supported by artificial intelligence	Number of Processes	All departments	Every 6 months

Source: own study

Findings and Recommendations

The main reason for creating Business Models 4.0 is to offer unique added value to the customer in the most profitable way possible for a company. Therefore, Business Model 4.0 should encourage the cooperation of all participants in the value chain, including motivating customers to be loyal to the supplier.

Hustinx, P. (2010). Privacy by design: Delivering the promises. *Identity in the Information Society*, *3*(2), 253–255.

Li, F., Nucciarelli, A., Roden, S., & Graham, G. (2016). How smart cities transform operations models: A new research agenda for operations management in the digital economy. *Production Planning & Control*, *27*(6), 514–528. doi: 10.1080/09537287.2016.1147096

Parmenter, D. (2015). *Kluczowe wskaźniki efektywności (KPI). Tworzenie, wdrażanie i stosowanie*. Helion.

Prahalad, C., & Krishnan, M. (2010). *The new age of innovation*. McGraw-Hill.

Saniuk, S., Saniuk, A., & Cagáňová, D. (2019). Cyber industry networks as an environment of the industry 4.0 implementation. *Wireless Networks*, 1–7.

Santos, J., Tarrit, K., & Mirakhorli, M. (2017). *A catalog of security architecture weaknesses*. https://design.se.rit.edu/papers/cawe-paper.pdf

Suomala, P., Sievänen, M., & Paranko, J. (2002). Customization of capital goods—implications for after sales. In *Moving into mass customization*. Springer.

4 Challenges of Industry 4.0 Development From Perspective of Supplier of Automating Production Process Solutions

Mirosław Moroz

Introduction

In the European Union, industry is responsible for 19.5% of gross value added (Eurostat, 2019). Thus, industry has the largest share of GVA generation. Industry is a flywheel for economic development, creating innovation, jobs, and exports. The concept of Industry 4.0 serves for the further development of the sector because its implementation in the daily operations of industrial enterprises allows them to increase production precision, reduce production time, and increase the flexibility of the production process, etc.(Moeuf et al., 2018).

The concept of Industry 4.0 (I4.0) is relatively new, as it was formulated in 2011 as part of an initiative launched by the German government. Since then, there have been numerous implementations of the Industry 4.0 concept, but not all of them have been successful. The implementation of the concept in question is not easy due to its interdisciplinary nature (technical, organizational, personnel, IT, financial, and logistical aspects). The challenges to and barriers from implementing the Industry 4.0 concept appear in almost every study. However, a literature review has shown that the research directly focused on identifying, classifying, and solving problems with the implementation of the I4.0 concept is in the minority. Research in this area was based on theoretical considerations or research in companies using Industry 4.0.

The author decided to look at the challenges of implementing the Industry 4.0 concept from the perspective of the manufacturer of solutions automating their production. Based on the case study method, the study identifies the problems related to the implementation of the Industry 4.0 concept identified during the design, implementation, installation, and servicing of automated production workstations.

This chapter is structured as follows. The first part contains an overview of the literature related to the Industry 4.0 concept, the components of this concept, and the advantages of its implementation. The next section presents the identified research related to the challenges

DOI: 10.4324/9781003186373-4

of implementing the Industry 4.0 concept (research gap) as well as the methodology (research purpose and problem, research methods, etc.). The challenges identified from the perspective of the solution provider of this concept during the implementation of the Industry 4.0 concept are the topics addressed in the next section, while recommendations for companies implementing the Industry 4.0 concept as well as the conclusions make up the final part of this study.

Literature Review

Concept of Industry 4.0

The term Industry 4.0 was publicly used at the Hanover Fair in 2011 (Xu et al., 2018). This was a sign of growing expectations for the application of ICT in the industrial sector. The possibilities of the economic use of the Internet were manifested to a large extent in applications related to customer service, information, and placing orders. E-commerce and e-marketing included a growing number of users, generating more and more revenue every year. On the other hand, there has been technological progress, both in the area of Internet access technology (wireless access, development of communication standards, increasing network capacity, etc.) as well as in the field of software automation and algorithmization. This trend referred to those solutions used in industrial machinery and equipment. Controllers and actuators have increasingly become computer-controlled and, to a lesser and lesser extent, mechanical or electronic solutions (Xu et al., 2018). In addition, ways of visualizing data and information processed by IT systems have been developed. The human-machine interface has included sophisticated methods such as augmented reality or virtual reality (Esengün & İnce, 2018). All of the factors mentioned above together have led to real possibilities for implementing the Industry 4.0 concept. As a result, interest in this concept has been growing, both among managers and researchers.

The term Industry 4.0 itself is not clearly understood in the literature; there are two reasons for this. First, this concept is based on an interdisciplinary approach, which refers to information, electronic, mechanical, psychological, organizational, and financial aspects. Second, individual industries are characterized by considerable specificity, which affects the scope and ways of implementing the Industry 4.0 concept. Table 4.1 provides a set of definitions of the Industry 4.0 concept.

The presented definitions indicate the three most important features of the Industry 4.0 concept. First of all, this concept was born from the technical possibilities of integrating systems, solutions, IT, and physical applications. Second, the cited definitions indicate which components make up the concept (the Industrial Internet of Things, Cyber-Physical Systems, Smart Factory). Third, as the concept matures, its boundaries

Table 4.1 Definitions of term "Industry 4.0" in the literature of the field

Author (year)	*Definition*
Kagermann et al. (2013)	"In essence, Industrie 4.0 will involve the technical integration of CPS into manufacturing and logistics and the use of the Internet of Things and Services in industrial processes. This will have implications for value creation, business models, downstream services, and work organization."
Lasi et al. (2014)	"The term Industry 4.0 collectively refers to a wide range of current concepts, whose clear classification concerning a discipline as well as their precise distinction is not possible in individual cases. [. . .] The concepts are: smart factory [. . .], cyber-physical systems [. . .], self-organization [. . .], new systems in distribution and procurement [. . .], new systems in the development of products and services [. . .], adaptation to human needs, and corporate social responsibility [. . .]."
Schmidt et al. (2015)	"In this paper, Industry 4.0 shall be defined as the embedding of smart products into digital and physical processes. Digital and physical processes interact with each other and cross geographical and organizational borders."
Hermann et al. (2016)	"The convergence of industrial production and information and communication technologies, called Industrie 4.0 [. . .]"; "In this publication, the authors name three key components of Industrie 4.0: the IoT, Cyber-Physical Systems (CPS), and Smart Factories."
Kang et al. (2016)	"Industry 4.0, or Smart Manufacturing, is the fourth industrial revolution. It is a new paradigm and convergence of cutting-edge ICT and manufacturing technologies. It provides ground for making effective and optimized decisions through swifter and more accurate decision-making processes."
Xu et al. (2018)	"Industry 4.0 has emerged as a promising technology framework used for integrating and extending manufacturing processes at both intra- and inter-organizational levels."
Aceto et al. (2019)	"The key concept behind I4.0 is integration, seen along three different axes: (i) horizontal integration, which regards cooperation between enterprises along a value chain; (ii) vertical integration, which refers to extensive automation inside the single enterprise; and finally (iii) end-to-end integration, which envisions connections across the value chains (realizing the value network) between possibly every couple of digitally enabled participants (machine-to-machine, human-to-machine, human-to-human)"

Source: Aceto et al., 2019, p. 3467; Hermann et al., 2016, p. 3928; Kagermann et al., 2013, p. 15; Kang et al., 2016, p. 119; Lasi et al., 2014, p. 240; Schmidt et al., 2015, p. 19; Xu et al., 2018, p. 2942

move toward the entire value-creation process, with an emphasis on the automation and autonomy of activities.

Components and Benefits of Industry 4.0 Concept

The Industry 4.0 concept is based on several components; the most important of these include Davis et al. (2012), Hofmann and Rüsch (2017), Lu (2017), and Alcácer and Cruz-Machado (2019):

1. The Industrial Internet of Things—connected sensors, controllers, and other industrial devices or machines as well as industrial computer applications that enable data collection, exchange, and processing in real time.
2. Cyber-physical systems—systems that consist of an integrated physical layer (machines, devices, actuators, etc.) and an IT layer (from data processing, analysis, and control of ongoing processes up to making autonomous decisions based on algorithms).
3. Big data and Predictive Analytics—processing large data sets, searching for patterns, and creating predictive models.
4. Autonomous robots—robots or devices that can perform actions for extended periods without human intervention.
5. Cloud computing—providing sufficient computing power; also, the results of data processing are available to every authorized entity via the Internet.
6. Human-machine interaction—allowing us to control and visualize the data obtained from devices or machines.

The correct implementation of the Industry 4.0 concept has the following advantages (Moeuf et al., 2018):

- Increased process flexibility;
- Reduced costs;
- Improved productivity;
- Improved quality of products;
- Reduced delivery times.

The factors that generate benefits can be classified in four areas (Ben-Daya et al., 2019; Moeuf et al., 2018; Zhong et al., 2017):

1. Monitoring:
    a. Product condition information;
    b. Machine condition information;
    c. Real-time information capturing;
    d. Enhanced interpretation for users.

2. Control:
   a. Visibility and traceability framework;
   b. Smart design and production control;
   c. Real-time production performance analysis;
3. Optimization:
   a. Highly modular production lines;
   b. Sharing data between applications;
   c. Customization of mass-produced parts.
4. Autonomy:
   a. Autonomous manufacturing;
   b. Networked and real-time decisions based on adaptative algorithms;
   c. Predictive maintenance using data mining and smart algorithms;

Research Methodology

Research Gap

Nearly every article devoted to Industry 4.0 refers to the challenges and barriers related to the implementation of this concept. However, a literature review revealed a relatively small number of studies presenting research that directly addresses Industry 4.0's challenges. The methodology of the conducted research is presented in Table 4.2, while the most important results of the mentioned research are presented in Table 4.3.

A relatively small number of studies on the identification and classification of challenges in the implementation of I4.0 indicate the existence of a research gap. Moreover, the study was conducted among managers whose companies are at various stages of implementing the I4.0 concept. A literature query did not reveal any research about the challenges of the Industry 4.0 concept from the perspective of other organizations involved in the I4.0 implementation process.

The following premises lead to such a conclusion:

- The supplier has developed competence in the I4.0 concept (substantive proficiency);
- The supplier not only participates in the implementation process but also maintains I4.0 solutions (long-term assessment of the actual effects of the implementation);
- Greater objectivity of assessing results "from the outside" (people who are not responsible for the implementation process looking at the implementation problems);
- Many implementations of the Industry 4.0 concept by the supplier (industry/time diversity).

Table 4.2 Research on challenges of implementing Industry 4.0 concept—methodological aspects

Author(s)	*Year of publication*	*Research method*	*Number of participants/ companies*	*Type of participants*
Kiel et al.	2017	Multiple case study	46	46 managers. Germany based
Kamble et al.	2018	Qualitative interviews	14	5 academicians, 3 IT engineers, 6 managers, Indian industry
Schneider	2018	Quantitative survey	130	44 academicians, 86 practitioners
Hamzeh et al.	2018	Quantitative survey	43	43 industrial practitioners. New Zealand industry

Source: Kiel et al., 2017; Kamble et al., 2018; Schneider, 2018; Hamzeh et al., 2018

Table 4.3 Identified areas of challenges and major problems

Author(s)	*Identified areas of challenges*
Kiel et al.	Technical integration Organizational transformation Data security Competition Cooperation Future viability Financial resources & profitability Human resources
Kamble et al.	Employment disruptions High implementation costs Organizational and process changes Need for enhanced skills Lack of knowledge-management systems Lack of clear comprehension about IoT benefits Lack of standards and reference architecture Lack of Internet coverage and IT facilities Security and privacy Issues Seamless integration and compatibility issues Regulatory compliance issues Legal and contractual uncertainty
Schneider	Analysis and strategy Planning and implementation Cooperation and networks Business models Human resources Change and leadership

(*Continued*)

Table 4.3 (Continued)

Author(s)	*Identified areas of challenges*
Hamzeh et al.	Additional investment of funds Additional investment of time Lack of access to people with the right skills Availability of equipment or software

Source: Kiel et al., 2017; Kamble et al., 2018; Schneider, 2018; Hamzeh et al., 2018

Research Problem, Purpose of the Study, and Research Method

The identified research gap has led to the following research problem: the assessment of awareness, staff, and organizational preparedness and the resources of companies implementing the Industry 4.0 concept from the perspective of providers of solutions that automate/robotize production processes.

The purpose of this chapter is to assess the challenges associated with the implementation of the concept of Industry 4.0 as perceived by the supplier of solutions that automate production process/Industry 4.0 solutions.

The research method used in this study is a case study. The case study method allows one to deeply penetrate the reality (context) of a given research object and capture the research problem in all of its complexity. This is important in the case of contemporary research problems, the knowledge of which is at an early stage of development (Dubé & Paré, 2003). Second, the case study method is advisable for studying the interdisciplinary issues (both social and those related to the use of information systems) that are challenges to implementing the I4.0 concept (Mettler & Rohner, 2009; Oesterreich & Teuteberg, 2016).

The research techniques used for the study include semi-structured interviews and observations. The interviews were conducted with two managers of the chosen supplier company: the CEO and the production manager. Such a selection of study participants resulted from the desire to receive comprehensive answers related to the full scope of activities in the process of implementing the I4.0 concept and as well as to take advantage of their many years of experience and competence. In turn, the observations covered the view of the possibilities of robotic integrated workstations and laboratories at ZAP-Robotics Ltd.

The research was conducted in December of 2019.

Brief Description of ZAP-Robotyka Ltd.

ZAP-Robotyka Ltd. designs and manufactures automated production systems. The history of the company dates back to the 1970s.

The company originated from the Industrial Automation Machinery Plant (ZAP) in Ostrow Wielkopolski, Poland. During the course of organizational and ownership changes, ZAP's company was divided into several related companies; one of these is the examined ZAP-Robotyka Ltd.

The company's activities include design, construction and testing, assembly and start-up, technical implementation, and the training and servicing of solutions. By 2019, the company implemented around 300 robotic and automated solutions in various industries (e.g., automotive, household products, glassware, and power engineering). Until 2010, the company produced IRp-6 and IRp-60 industrial robots; nowadays, the company's output is based on Fanuc's robots. The company is working closely with two universities to implement I4.0 projects. The first implementations of the I4.0 concept took place in 2016. Since then, the company's staff has participated in 10 implementations of the I4.0 concept, constantly improving their competence in this area. ZAP-Robotyka Ltd. is a small company with 45 employees. The implementation of the Industry 4.0 concept has taken place in companies that are integrated into the global supply chain. These are manufacturers of components for household appliances as well as for the automotive and machine industries. Thus, their products ultimately reach the global market even though the capital of the above-mentioned industries was formally Polish.

Study Results: Identified Challenges During Implementation of Industry 4.0 Concept

Challenges and barriers to the implementation of Industry 4.0 in the opinion of the managers of a company that provides automated/robotized production systems

The interview with both the CEO and the Production Director of ZAP-Robotyka Ltd. allowed forth identification of the main challenges for the implementation of the Industry 4.0 concept on the part of the companies implementing the concept. From the perspective of an I4.0 solution provider, the most frequent challenges of the enterprises that implement the said concept are as follows:

1. Insufficient level of preparation of company's personnel (in particular, engineering and maintenance staff);
2. Deficiencies in IT infrastructure;
3. Significant capital needs related to launching implementation;
4. Failure to prepare partners of companies implementing Industry 4.0 concept for accuracy of parts and components;
5. General views of companies implementing I4.0 on maintenance issues in automated/robotic/integrated production systems.

Ad 1.

Production processes are a complex system of interrelated material elements (machines, devices, equipment), human (skills, commitment, openness to change, organizational cultures), and organizational aspects (rules of position grouping, responsibilities, and activities, division into changes). The implementation of automated systems within the I4.0 concept results in major changes in each section. The staff directly involved in the production processes (engineers, technicians) are most affected by the change related to the implementation of an I4.0 production system. What had previously required effort and attention started to happen on its own from that moment. Machines integrated into the system with their own control software pick up parts, weld, arrange for further transport, etc. Of course, people must supervise and monitor the production processes in terms of maintaining production continuity, having sufficient inventory, and responding to unusual situations. This means a change in the basic responsibilities for the direct production staff.

It involves moving from responsibilities such as imposing elements, processing evaluation, making decisions regarding the quality of workmanship, and then the fate of the manufactured detail, to obligations related to monitoring the production process and responding to critical situations. Theoretically, it does not seem that getting rid of physical activities or portions of the evaluation and decision-making duties would be problematic; however, the experience of the ZAP-Robotics Ltd. managers indicates that human resources are a critical factor in the success of the implementation. Current habits, disbelief in the adaptability of I4.0 systems and their ability to cope with unusual situations, and finally conservatism and reluctance to change usually result in an indifferent attitude toward the implemented solutions. And this, in turn, extends the time of the individual stages of implementing an I4.0 system (design, assembly, commissioning). So, from this point of view, the key is to prepare the direct production staff for changes—to mentally prepare them for openness to change. It is worth emphasizing at this point that the obstacle in implementation was not insufficient expertise nor the insufficient technical preparation of the engineers or technicians.

Ad 2.

Implementation of the I4.0 concept involves a large amount of transmitted data. For example, the implementation of a Class I4.0 system by integrating nine vision sensor robots with software that ensures production autonomy and the full traceability of the manufactured components generates the need for the continuous transmission of significant data streams among all components of the system. Very often, companies implementing the I4.0 concept have not prepared sufficient IT resources—adequate server capacity nor sufficient internal network capacity. The indicated barrier is revealed during the installation and start-up phase of the system. Since the challenge is purely technical, it can be eliminated relatively quickly by obtaining the correct server.

Ad 3.

The implementation of the Industry 4.0 concept requires companies that are willing to set aside substantial budgets. Due to the high individualization of each I4.0 system implementation project (the scope of activities, number of robots, number of sensors, types of sensors, complexity of the problem of ensuring precision and repeatability of operations, etc.), it is not possible to determine a uniform price list. Nevertheless, according to the estimates of ZAP-Robotyka Ltd.'s managers, the average cost of implementing the I4.0 system is about 1 million PLN (around 265,000 USD). At this point, it is worth noting that the cost of a comparable I4.0 system is about three–four times greater in the countries of Western Europe or North America. For those companies operating on the market, the implementation of I4.0 solutions is an investment that should pay off. And here the challenge arises—the return on such an investment is only possible in a situation of mass production or long-term production (which, however, does not exclude the production of many product versions). It is not possible for many companies or industries to meet these conditions. In this context, it is not surprising that the automotive or household appliance industries are at the forefront of implementing the Industry 4.0 concept. This is because they are characterized by long production series aimed at a wide range of retail customers. On the other hand, it is worth noting that the labor costs are increasing in Central and Eastern European countries while the gap between the unit costs of labor production and automated systems is decreasing. This likely results in the fact that new industries are becoming interested in implementing the I4.0 concept. Nevertheless, based on the experience of the ZAP-Robotyka Ltd.'s managers, the need to invest a significant amount of money initially deters those companies concerned. This challenge is difficult to overcome for a single organization because it is related to objective operating conditions (scale of activity, nature of production, or industry specificity).

Ad 4.

The possibilities of making full use of the advantages of Industry 4.0 systems are determined by the accuracy of the components or subassemblies to be processed in the factory of a company implementing the I4.0 concept. Automated systems work properly when dimensional tolerances are maintained; however, it is worth noting that the precision of the workmanship must be greater than it was before when the component was manually processed. The Industry 4.0 class system therefore requires changes not only in the company that implements the system but also in its partners. Unfortunately, there are problems with maintaining the precision and full repeatability of the components. According to the observations of the ZAP-Robotyka Ltd.'s managers, this is a challenge that arises during the assembly and start-up phases. It is not easy to change the current production settings of a cooperating company nor the habits of the staff. Sometimes, an existing supplier changes if no agreement can be reached. Thus, it is necessary to inform the cooperating party in advance about the need to raise the standards of precision and repeatability of the delivered parts for the effective implementation of the I4.0 concept.

Ad 5.

Another challenge that has been observed by the managers of ZAP-Robotics Ltd. in the process of implementing the I4.0 concept is the misperception of maintenance services about the functioning of I4.0 solutions. By their very nature, I4.0-class systems can operate non-stop. This involves the need to work in two or three shifts. De facto, therefore, the work is continuous, with all of the consequences. In this context, the challenges are to remodel the work organization so that the I4.0 system is under constant supervision. In the case of previous one-shift work, this raises larger or smaller conflicts between the employer and employee as well as between the employees themselves. Second, the maintenance staff should be made aware of the need for the periodic maintenance of an I4.0 system as well as any emergency response. In particular, this is about having the most sensitive parts in storage and training a team to fix basic faults so that they can carry out uncomplicated repairs themselves without prolonged downtime. This type of challenge appears at the stage of operation (service).

The questioned managers of ZAP-Robotics Ltd. also determined the level of experiencing specific challenges among enterprises that decide to use I4.0 systems. Table 4.4 presents the above issue.

The challenge related to the attitudes of production staff (openness to changes) is perceived as the strongest.

The gap between the automotive and household appliance industries and other sectors of the economy is also visible; they experience far fewer challenges when implementing I4.0 solutions.

Recommendations for Enterprises Implementing Industry 4.0 Concept

Directly, production employees have concerns about what their work will look like—their responsibilities after the I4.0 system is launched. They are also afraid of their jobs being reduced. The implementing

Table 4.4 Level of experiencing challenges in assessment of I4.0 solution provider

Challenge	*Automotive, household appliances*	*Other industries*
High cost of implementation	1	2.5
Lack of understanding of Industry 4.0 concept	1	1
Attitude of production staff (responding to changes)	2	3
Need to change technological process	1	2
Deficiencies in managerial skills	1	1
No vision for implementing I4.0 concept	1	1
IT concerns	1	1.5

Source: author' depiction.

Note: 1—no noticeable challenges; 3—major noticeable challenges.

company should better communicate the scope of the changes, pointing out the advantages of automation such as the transition from physical work to work related primarily to monitoring the condition of the robot and improving health and safety conditions (e.g., during welding, the work is less harmful to the eyes and lungs). In order to increase the esprit de corps, companies should have a sound information policy on employment, indicating this in general terms of the changes of duties, not the number of posts.

Similarly, a company implementing I4.0 should pay attention to changing the habits and attitudes of the maintenance staff. From an operational point of view, it is this group of employees who will be responsible for ensuring the continuity of production. It is worth working on changing the attitude to multi-shift work as well as providing substantive training related to monitoring the I4.0 system and replacing typical system elements.

Another suggested recommendation is to disseminate information about the implementation of the I4.0 concept across the entire supply chain. It is worthwhile to inform the cooperating companies in advance that as a result of implementing I4.0 systems, it is necessary to increase the accuracy of the production components. It is important here to communicate precise requirements for future dimensional tolerances as well as allow for sufficient time to actually implement more precise production.

As a problem of an IT nature, there was an issue of insufficient disk space and too-low network throughput. However, it seems that cloud computing (with adequate cybersecurity) and more efficient IoT networks should be deployed.

Conclusions

Based on the study, it can be concluded that the awareness of the existence, advantages, and disadvantages of the I4.0 concept is high among company management as related to enterprises implementing the Industry 4.0 concept through ZAP-Robotyka Ltd. It follows that the I4.0 concept has become common among those industrial enterprises interested in automating production.

Preparing the recipients of I4.0 systems looks slightly different. From the perspective of the supplier of automated systems, the need for better preparation in terms of personnel, organization, finances, and technology is noticeable. Taking into account the frequency of occurrence and weight together, the most serious challenges are personnel and organizational issues, as they mostly deal with issues on which the implementing company has an impact. There is the real influence of management and the opportunity to more fully prepare for the implementation of the I4.0 concept.

From the point of view of resources, a noticeable challenge is finding free financial resources for investments in I4.0 systems. This is undoubtedly a barrier to implementation, but this barrier is not related to the specificity of the I4.0 concept—it is appropriate for all investments whose results are postponed (Schmalensee, 1981).

The quality of the engineering and technical staff is most responsible for the effectiveness of the implementation—their competencies and attitudes are a key challenge for companies implementing the Industry 4.0 concept according to the observations of the managers of ZAP-Robotics Ltd. In the end, this is positive information, because the competencies and attitudes of employees can be shaped by the company (although it is not necessarily an easy process).

The primary condition for implementing the Industry 4.0 concept is the industry specificity associated with the length of the production series. In industries such as automotive and home appliances, there are definitely smaller resource and organizational-personnel barriers. Other industries and smaller enterprises face greater challenges in implementing the Industry 4.0 concept. The above statement finds support in the literature. Research conducted among German small- and medium-sized enterprises showed their smaller adaptation to the implementation of the I4.0 concept (Sommer, 2015). Based on the above research, the level of challenges can be classified due to the size of the organization and the subject of the activity. Figure 4.1 reflects the above idea.

The relatively high costs of implementing I4.0-class systems result in the fact that the implementation may be decided by those companies for which the difference between the unit cost of production by the employee and the automated system will be sufficiently low. This is confirmed by the observations of the ZAP-Robotyka Ltd.'s managers as well as data on work performance in Asia. Adidas has set up two factories in Germany

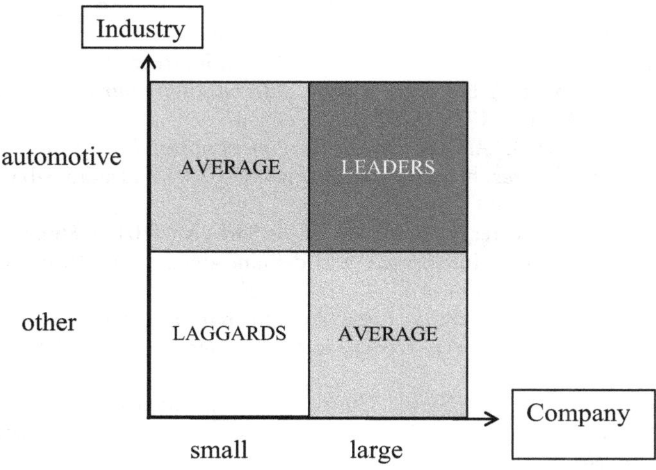

Figure 4.1 Level of perceived challenges due to company size and the subject of activity

Source: author's depiction

and the United States according to the I4.0 concept. However, labor costs were so low and the competences of the local workers were so high in Asian countries that further production became unprofitable, so the factories closed down after three years (Coldewey, 2019). This leads to the thesis that the boundary condition for the dissemination of Industry 4.0 solutions will be the decreasing difference between the level of salaries and the costs of implementing the I4.0 concept.

The above studies have limitations. Conclusions and recommendations were generated on the basis of case study analysis and literature review. It would be worthwhile to verify the results and conclusions of the study by conducting a survey on a larger/representative sample of companies. Another limitation is related to conducting research among enterprises located in Poland. While they have been integrated into the global supply chain, the attitudes and skills of the crews of those companies implementing I4.0 concepts are influenced by homogeneous social and cultural factors. In the future, it would be worth undertaking intercultural/international research.

References

Aceto, G., Persico, V., & Pescapé, A. (2019). A survey on information and communication technologies for industry 4.0: State-of-the-art, taxonomies, perspectives, and challenges. *IEEE Communications Surveys & Tutorials*, *21*(4), 3467–3501.

Alcácer, V., & Cruz-Machado, V. (2019). Scanning the industry 4.0: A literature review on technologies for manufacturing systems. *Engineering Science and Technology, an International Journal*, *22*(3), 899–919.

Ben-Daya, M., Hassini, E., & Bahroun, Z. (2019). Internet of things and supply chain management: A literature review. *International Journal of Production Research*, *57*(15–16), 4719–4742.

Coldewey, D. (2019). *Adidas backpedals on robotic shoe production with speed factory closures*. https://techcrunch.com/2019/11/11/adidas-backpedals-on-robotic-factories/

Davis, J., Edgar, T., Porter, J., Bernaden, J., & Sarli, M. (2012). Smart manufacturing, manufacturing intelligence and demand-dynamic performance. *Computers & Chemical Engineering*, *47*, 145–156.

Dubé, L., & Paré, G. (2003). Rigor in information systems positivist case research: Current practices, trends, and recommendations. *MIS Quarterly*, *27*(4), 597–636.

Esengün, M., & İnce, G. (2018). The role of augmented reality in the age of industry 4.0. In A. Ustundag & E. Cevikcan (Eds.), *Industry 4.0: Managing the digital transformation* (pp. 201–215). Springer.

Eurostat. (2019). *National accounts and GDP*. https://ec.europa.eu/eurostat/statistics-explained/index.php?title=File:Gross_value_added_at_current_basic_prices,_2008_and_2018_(%25_share_of_total_gross_value_added)_FP19.png

Hamzeh, R., Zhong, R., & Xu, X. W. (2018). A survey study on industry 4.0 for New Zealand manufacturing. *Procedia Manufacturing*, *26*, 49–57.

Hermann, M., Pentek, T., & Otto, B. (2016). *Design principles for Industrie 4.0 scenarios*. 2016 49th Hawaii International Conference on System Sciences (HICSS), pp. 3928–3937. IEEE.

Hofmann, E., & Rüsch, M. (2017). Industry 4.0 and the current status as well as future prospects on logistics. *Computers in Industry*, *89*, 23–34.

Kagermann, H., Wahlster, W., & Helbig, J. (2013). *Recommendations for implementing the strategic initiative Industrie 4.0: Final report of the Industrie 4.0 Working Group*. https://docplayer.net/254711-Securing-the-future-of-german-manufacturing-industry-recommendations-for-implementing-the-strategic-initiative-industrie-4-0.html

Kamble, S. S., Gunasekaran, A., & Sharma, R. (2018). Analysis of the driving and dependence power of barriers to adopt industry 4.0 in Indian manufacturing industry. *Computers in Industry*, *101*, 107–119.

Kang, H. S., Lee, J. Y., Choi, S., Kim, H., Park, J. H., Son, J. Y., Kim, B. H, & Noh, S. D. (2016). Smart manufacturing: Past research, present findings, and future directions. *International Journal of Precision Engineering and Manufacturing-Green Technology*, *3*, 111–128.

Kiel, D., Müller, J. M., Arnold, C., & Voigt, K. I. (2017). Sustainable industrial value creation: Benefits and challenges of industry 4.0. *International Journal of Innovation Management*, *21*(8), 1–21.

Lasi, H., Kemper, H-G., Fettke, P., Feld, T., & Hoffmann, M. (2014). Industry 4.0. *Business & Information Systems Engineering*, *6*(4), 239–242.

Lu, Y. (2017). Industry 4.0: A survey on technologies, applications and open research issues. *Journal of Industrial Information Integration*, *6*, 1–10.

Mettler, T., & Rohner, P. (2009). Supplier relationship management: A case study in the context of health care. *Journal of Theoretical and Applied Electronic Commerce Research*, *4*(3), 58–71.

Moeuf, A., Pellerin, R., Lamouri, S., Tamayo-Giraldo, S., & Barbaray, R. (2018). The industrial management of SMEs in the era of industry 4.0. *International Journal of Production Research*, *56*(3), 1118–1136.

Oesterreich, T. D., & Teuteberg, F. (2016). Understanding the implications of digitisation and automation in the context of industry 4.0: A triangulation approach and elements of a research agenda for the construction industry. *Computers in Industry*, *83*, 121–139.

Schmalensee, R. (1981). Economies of scale and barriers to entry. *Journal of Political Economy*, *89*(6), 1228–1238.

Schmidt, R., Möhring, M., Härting, R. C., Reichstein, C., Neumaier, P., & Jozinovic, P. (2015). Industry 4.0-potentials for creating smart products: Empirical research results. In W. Abramowicz (Ed.), *Business Information Systems* (pp. 16–27). Springer.

Schneider, P. (2018). Managerial challenges of Industry 4.0: An empirically backed research agenda for a nascent field. *Review of Managerial Science*, *12*(3), 803–848.

Sommer, L. (2015). Industrial revolution-industry 4.0: Are German manufacturing SMEs the first victims of this revolution? *Journal of Industrial Engineering and Management*, *8*(5), 1512–1532.

Xu, L. D., Xu, E. L., & Li, L. (2018). Industry 4.0: State of the art and future trends. *International Journal of Production Research*, *56*(8), 2941–2962.

Zhong, R. Y., Xu, X., Klotz, E., & Newman, S. T. (2017). Intelligent manufacturing in the context of industry 4.0: A review. *Engineering*, *3*(5), 616–630.

5 Strategic Information Analytics System in Context of Challenges of 4.0 Age

Michał Baran

Introduction

The contemporary globalized economy is undergoing constant and dynamic changes. These changes occur primarily under the influence of crucial technological solutions that primarily concern the sphere of the exploitation of information. As a consequence, the existing rules of economic exchange are rapidly evolving, which shifts the relative bargaining positions of individual links in the supply chain. Some of these links are gaining importance, while others are losing it. The current distribution of forces between organizations that create a complicated system of interrelated entities is disturbed. At the same time, it is often difficult to predict the long-term consequences of even minor modifications to the existing architecture of the entire system of related elements.

Viewed from the perspective of a single corporation, the speed of the changes taking place requires the constant conducting of extremely careful analysis that enables an understanding of the essence of the ongoing revolution and the directions of its further course. What is at stake is the possibility of constantly identifying new spaces that create the opportunity to generate significant added value (and, thus, achieve satisfactory profits). In the largest economic organizations operating under the pressure of global competitors, the strategic information analytics system mentioned in the title (defined in detail later in the discussion) becomes a key element in the area of management. This is because of the role that effective access to properly processed information begins to play, it being followed by the transmission of this processed information to those people who are responsible for making strategic choices. The implementation of these types of tasks by traditional systems with limited bandwidth and range of operation encounters significant barriers in situations where the level of the saturation of data sources observed in the environment (even if potentially only within reach) causes that the problem of the excessive proportion of the data sources becomes more apparent. In addition, this data usually has a very diverse character and form of recording. Therefore, without the use of the appropriate tools, it cannot be easily interpreted due to the limitations of human perception.

DOI: 10.4324/9781003186373-5

There is a constant race among competing companies for being the first to gain the opportunity to recognize and also correctly interpret new trends and phenomena (not yet noticed by others), which can then be used to strengthen their own bargaining power. In this situation, the challenges arising from the growing level of complexity in the environment (in which economic exchange takes place) trigger a reaction in the form of the development of advanced system solutions in the sphere of information processing. Such solutions are systematically improved by professionally and consciously managed entities in order to enable them to effectively operate under the conditions set by the Fourth Industrial Revolution. Having the opportunity to conduct research on a group of the largest Polish corporations, the author asked about the levels of readiness of these entities to effectively carry out the necessary tasks arising from the occurrence of the situation described earlier. In light of the results of the conducted analyses, it has been possible to diagnose a number of interesting relationships, thereby shifting the horizon of the available knowledge on the strategic aspects of the functioning of the largest enterprises that operate under increasing pressure from global competition and a dynamically changing environment.

Determinants of Strategic Information Processing Related to Fourth Industrial Revolution

The development of the digital economy has caused a dynamic change in the arrangement of the sources of competitive preponderance that the vast majority of large enterprises competing on international markets got used to over the years (when this arrangement remained relatively stable) (Antonelli, 2014). The revolution that is taking place nowadays results in the necessity of much more intense competition for access to information because, in most areas today, this access is absolutely crucial. Only because of having a properly built and constantly expanded information set is it possible to pave the way for success in the coming years (Van Rijmenam, 2019). However, it should be remembered that, in addition to obtaining information, parallel competencies should be developed in the sphere of the effective processing of this information—both in terms of procedure and IT technology (Becker, Knackstedt, Pöppelbuß, 2009). To some extent, this can be done by improving the solutions used within the work organization system of the staff employed by an enterprise (Dziekoński, 2017). However, it is more and more often no longer a human being—the one who becomes a sender and recipient of information—but this role belongs to a machine, as the popularity of using artificial intelligence is growing rapidly in many areas of life (Zhou, 2013). The range of the applications of this technology is becoming broader—it can be seen even when observing the implementation of various, even small autonomous devices accompanying humans in everyday life.

Nowadays, technologies based on the Internet of Things and Services are becoming more common, which leads to the generation of a complex system with a decentralized structure and a very dense network of interdependencies and mutual conditioning (Manu, 2015). These types of connections operate in the form of feedback loops, setting a completely new level of complexity in the world created in this way. Both in the case of physical and office work, robotization and automation are progressing rapidly. These phenomena are accompanied by the autonomy of technological lines, processes, and individual devices. The human factor ceases to play its current role (Larsson & Teigland, 2019). The disappearance of a human presence is particularly evident at the operational level—where technologically advanced forms of doing business are introduced. The saturation of the technosphere with sensors operating in a feedback loop in the process of predicting and forecasting changes in the environment allows for a more and more flexible response to any new circumstances as well as a preventive avoidance of errors and mistakes. The phenomenon mentioned above usually takes place in real time. The effect of these changes is the emergence of a new form of business organization in the form of so-called cyber-physical systems for the production of goods and the delivery of services (Monostori, 2014). In many cases, we are dealing with a so-called intelligent product; that is, a situation when suppliers collect feedback information from a product (which is also a sensor), which takes place after the moment in which the product is given to the user. Data on what happens to a product when it remains at the client's disposal is analyzed in order to improve the goods and services offered by the organization—so, the mechanism of entangling the client in creating value makes him one of the key participants in this process (Trstenjak & Cosic, 2017). There are also solutions that use the so-called incremental logic, which means giving the final shape of the goods and services delivered only after some time (Lanza et al., 2015). Their character gradually develops as the scenario of subsequent events develops under the influence of actions taken by the user.

The elimination of the importance of two basic barriers in the form of space and time allows one to effectively combine the spheres of design, production, sales, and after-sales service into one effective and fully optimized whole (Chen & Li, 2018). A company can build relationships with clients and other stakeholders on the basis of a smooth transition to the next iterations, anticipating their expectations and preparing to meet their future needs in advance (whose existence the monitored partners are not even aware of at the moment). The methodical approach allows one to build an optimized network of related entities according to the priorities of the cooperation initiator (Schuh et al., 2014). However, this also means blurring the existing boundaries between an organization and its environment. What occurs is the far-reaching mutual penetration of cooperating systems whose operation becomes impossible to be maintained at the correct level in the event of attempts to separate them again (Bukowski, 2016). In the new reality (which covers more and more areas of the functioning of both a single individual and entire organizations), the integration of previously independent elements within the smart idea is becoming commonplace. Self-regulating environments create friendly conditions for the functioning of a single consumer as well as entire artificial organisms (such as a transport network or an entire city). The phenomena of personalization and customization are progressing in all spheres of life, which the average user often accepts blindly by quickly adopting an offer addressed to him as a natural standard (Rautenstrauch et al., 2012).

Products and services based on augmented reality are starting to be implemented and commercialized in many areas. Such solutions allow us to enrich traditional products and services with additional functionalities and integrate them with offers of related goods and complementary services. Generally, it should be stated that the deepest changes affect the sphere of production; however, this sphere is defined very broadly today due to the degree of its connection with other business areas (Kagermann, 2014). For example, the way in which goods and services are consumed is now rapidly changing because, in many cases, there is a tendency for resignation from their possession in favor of payment for their use under solutions created by models based on the sharing economy (Strømmen-Bakhtiar & Vinogradov, 2020). This results in a change in the orientation of many enterprises from product to service. Due to various types of alliances, there follows an effect of the mutual strengthening of the effectiveness of the promotional and sales activities undertaken by organizations that decide to participate in a common data exchange system. However, it is necessary to have the right attitude of alliance-forming business entities, consisting of their being open to real cooperation and their sharing collected information, which makes them dependent on one another.

Nowadays, a significant number of solutions appear in the entire technological environment of a modern organization, which change the current rules of the game through the synergy effect. Substitute goods and services are being developed (compared to the offers currently available on the market) that are better suited to the needs and expectations of the final users. Multi-material 3D printers and other similar devices constituting a specialized element of an intelligent system (which is surrounded by an integrated information system network) eliminate the importance of the scale effect by depriving the existing leaders of their significance. In turn, virtual reality begins to be used to shorten many stages of product development (e.g., during its testing). Biotechnology, synthetic biology, nanotechnology, cryptology, and quantum computers result in solutions that are already innovations in themselves; only when they are combined into a larger whole, do they open completely new possibilities. Blockchain technology is becoming more widely used in financial markets as well as in logistics and reporting. Solutions based on this idea allow us to identify and track the history of changes while simultaneously precisely indicating the centers of responsibility. Blockchain gives one the chance to effectively secure all transactions by introducing a previously unreachable level of transparency of any interventions in the monitored reality. The order of subsequent entries that takes the time variable into account is a valuable source of extremely precise and useful knowledge that can be used in a practical way (inter alia, within the frame of cost-financial analysis but also in many other areas), thereby improving the operation of an entire complex system of related elements participating in registered transactions (Ryu & Won, 2018). When it comes to data security issues, another solution that is dynamically gaining popularity—despite some business concerns regarding safety questions and data control—is cloud computing (Kouatli, 2016). The main benefit here is recording all information in one system and making it available to all employees (as well as authorized partners) in order to create various types of complex analyses (regardless of when and where these employees and partners work).

The Fourth Industrial Revolution leads directly to solutions in which human perception turns out to be too limited to function independently in the new world that emerges in the form of the next stage of civilization development. Man begins to accept that artificial intelligence algorithms interpret complex data sets more effectively—however, such a solution also begins to arouse serious controversy due to the issue of responsibility for autonomously made decisions. A good example in this respect is the dynamic development of technologies allowing for the creation of system solutions in the field of autonomous transport and logistics (Bukowski, 2019). It seems that the limit for implementing innovative technologies to replace human beings is the level of superior universal values common to all humanity. Although it is possible to imagine the elimination of situations requiring the involvement of human resources for physical or intellectual work, moral and ethical issues are related to the conscious single human being (who bears the burden of responsibility)—this last sphere is out of the reach of even the most perfect machines.

All of the circumstances described earlier are not insignificant for the shape of a corporation's internal structure. The world of modern organizations requires a gradual resignation from the hierarchical vertical system. This traditional system reflects the division of power and is expressed in the form of command, control, and regulatory ties. Such a solution is increasingly being replaced by a system based on functional ties (expressing the division of labor) consistent with the course of horizontally oriented processes. This new model is based on cooperation founded on the opportunities created by dynamically developing mechanisms for effective information exchange. Organizational ties expressing the division of labor based on the exchange of information, cooperation, and mutual transfer of the effects of the implemented activities turn out to be crucial here. The growing importance of information ties must consequently lead to significant hierarchical changes, as democratized access to information means the democratization of power. It can also be argued that, in a turbulent environment, the only rational strategy is to focus the efforts of an organization and its partners even more strongly on achieving goals that are directly related to the core business. It should be strived for everything to follow the direction of the flow of an undisturbed stream of generation values (Tworek et al., 2019). However, it is important to base this collaboration on the use of a new level of information system. The manifestation of this process is the emergence of the so-called Teal Organizations, in which human resources are characterized by a high level of internal self-motivation actively cooperate while being based on the principle of universal partnership (Laloux, 2014). This is possible due to a properly created synthesis of personal goals of individual people with actions for the success of an entire organization.

The analysis conducted so far indicates the existence of three basic trends that affect the enterprise management model—a model that should be ready for the challenges of the 4.0 era. The first of these trends concerns the sphere of phenomena occurring in the environment of an organization and expresses the disappearance of the importance of the existing sources of competitive advantage. The second trend concerns the changes that take place inside the organization and leads to delegation to the outside of operational activities in order to concentrate all effort on taking strategic actions, which is accompanied by disruptive thinking (Schwab, 2017). The effect of the occurrence of the third key trend is the necessity of the continuous reconfiguration of a conducted activity—both in terms of the adopted internal solutions as well as the relationships established with the environment. In the case of large global corporations, this is particularly difficult due to the inertia that results from the need to effectively use previously existing resources (often of a very diverse nature). In this context, it becomes obvious that actions at the strategic level become even more important. The company's management must constantly perform three basic tasks. The first of them consists of producing the ad hoc design of a coherent and optimized sequence of activities, which go beyond the framework of the organization (and, thus, also include partners and users) and which are expected to generate added value. The second task is to monitor and constantly improve the activities in which the organization has already been involved by determining the appropriate parameters, procedures, and standards whose immediate interest is time-limited. Finally, the third key task of the management board is to decide on issues regarding the transfer of parts of the process to employees, robots, autonomous modules, subcontractors, etc. (however, the management board should take many specific issues into account such as security, costs, maintaining control over the added value, etc.). To be able to meet such challenges, the management needs support from an appropriate tool that guarantees that all necessary information (having a very heterogeneous nature due to the variability and differentiation of important circumstances) will be properly and timely analyzed and presented in a form that can be immediately used.

One possible solution that allows a corporation to meet the challenges of the Fourth Industrial Revolution is the creation of a strategic information analytics system (SIAS). The term SIAS will be understood as the advisory body of top-level managers who, each time, activate its operation by defining the need for individual analyses (constituting a direct basis for making the most important decisions). It is only a clearly identifiable and formally separated part of the organization whose functioning as a coherent whole is dedicated to fulfilling the role understood in this way (such a definition eliminates the consideration of cases of the hidden performance of these tasks in a dispersed and informal manner, which is difficult to grasp and unambiguously describe). The specificity of the strategic information analytics system as a generator of advanced expert studies allows us to demonstrate its unique identity situating it as a kind of extension of the MIS management information system. The separateness of SIAS results from the fact that it implements the unstructured tasks of a non-routine nature and of a low degree of repeatability, which are defined as part of a dynamic dialogue with the final user of analyses in the context of its current needs and expectations. The lack of a simple cyclical repeatability of the performed analyses (prepared in relation to non-standardized needs) is a feature that distinguishes SIAS from a strategic information system whose domain is to track the dynamics of selected phenomena of strategic importance on a continuous basis.

Methodological Conditioning for Conducting Research on Condition of Strategic Information Analytics Systems

In Poland, the group of enterprises included in the large-size category is composed of more than 5,000 entities. Among them are branches of large global corporations with formal legal independence that, in fact, do not make the most important strategic decisions alone (due to the lack of real autonomy as related to the foreign headquarters). Other entities that may be included in this group are also those companies operating in the space of regulated markets (e.g., the energy sector), where it is difficult to talk about implementing free competition rules. Pilot studies have allowed us to estimate the number of large entities that have actually developed clearly identifiable strategic information analytics systems in their structures (understood in accordance with the definition given in the previous point of consideration) per 1,000 cases. Interestingly, all of these companies have a turnover of at least €25 million, which may indicate that they are entities operating on a scale where global competition is strongly felt.

By starting to develop a research tool to enable us to test the condition of strategic information analytics systems in the identified group of corporations, the focus group method was used. After gathering small expert groups composed of employees employed in units that perform tasks assigned to a strategic information analytics system on a daily basis, they were asked questions about any problems they encountered at work. As a result of conducting this type of qualitative research, a set of five questions corresponding to the thesis adopted in the considerations was obtained. This thesis says that, during the Fourth Industrial Revolution, the most important challenge from the perspective of a company's success is the excessive data (which is also characterized by significant mutual differentiation) and the need to integrate it within one set as well as detecting their relationships and the accurate interpretation of the identified relationships. The set of questions contained the following elements:

- Q1 How often does it happen that you receive an excessive amount of information from various sources, making it impossible to analyze it holistically?
- Q2 Do you use any IT tools to combine data of a different nature or belonging to extensive data sets in such a way as to be able to draw conclusions resulting from the detection of previously unnoticed relationships?
- Q3 How often do you encounter organizational barriers in accessing the data needed to prepare a report, as the data is being held by other organizational units?
- Q4 How many changes are needed in the sphere of ways to get information (1 indicates completely unnecessary changes and 5 indicates absolutely necessary changes)?
- Q5 How many changes are needed in the sphere of data and information processing processes (1 indicates completely unnecessary changes and 5 indicates absolutely necessary changes)?

One-hundred-twelve large Polish enterprises from various industries participated in the study (it was checked that the variety parameter did not affect the results obtained) from the entire population of about a thousand units classified as a potential data source. Such a sample size means that, at a confidence level of 0.95 and while assuming a fraction of 0.2 (initially, it was assumed that 20% of the surveyed entities experienced 80% of the cases of a given problem in accordance with the W. Pareto rule), an estimation error may appear at a maximum of 7%. With a population of this size, obtaining better parameters would require a significant increase in the size of the research sample, which was considered to be unrealistic in the view of the attitudes of most enterprises. Only those employees who were directly involved in the operation of a strategic information analytics system were respondents. It has been assumed that the basis of inference will not only be the analysis of the distribution of responses obtained but also be the verification of any correlations that occur among them.

Results of Research Carried Out on State of Strategic Information Analytics Systems

Representatives of all 112 subjects participating in the survey answered each of the five questions asked. There was no significant impact of any additional variable (industry, turnover, dominant form of ownership) on the method of answering. The respondents did not report any difficulties regarding the possible ambiguity of a question asked nor other such doubts.

The first of the questions asked (Table 5.1, Part A) concerned the issue of possible problems that may be caused by information overload, resulting in the inability to analyze an entire collection. Nearly a quarter of the respondents reported frequent (19.64%) or very common (4.46%) difficulties in this respect. There were only 2.68% that indicated a problem when assessing this situation. The answers dominated that this problem is "rather rare" (44.64%), and the next option (when it comes to the frequency of indications) was the option "very rare" (28.57%).

The second question (Table 5.1, Part B) was to verify the scope of using IT tools to combine data of a diverse nature from a large set of data into coherent wholeness in order to detect previously unnoticed relationships. In the case of the surveyed group of the largest corporations with strategic information analytics systems, such tools are definitely used by 25.89% of the entities and 38.39% (the most frequently chosen answer) do it to a significant extent. Only one respondent did not take a clear position in this matter. In contrast, 25.89% of the respondents said that they rather do not use this type of support. The possibility of using IT tools in the implementation of the described task is unknown for 8.93% of the respondents.

The third question (Table 5.1, Part C) concerned the problem of organizational barriers in accessing the necessary data already in an organization's possession. Only 4.46% of the respondents pointed to the very frequent occurrence of such barriers, and 12.50% said that it happened often. No one had any difficulty assessing the situation in question unequivocally. The rather rare occurrence of difficulties of this nature was indicated most by as much as 44.46% of the respondents and 38.39% of them considered such a possibility to be very rare.

In the fourth question asked of the respondents (Table 5.1, Part D), they were asked to assess the need for change in the sphere of how to obtain the information they need. In all, 6.25% of the respondents gave the highest priority to this need, and 17.86% were slightly less critical in their assessment of the situation. Most people (31.25%) were moderate

Table 5.1 Distribution of answers to Questions: Q1, Q2, Q3, Q4, and Q5

Answer	*Quantity*	*Percentage*
Part A		
How often does it happen that you receive an excessive amount of information from various sources, making it impossible to analyze it holistically?		
Very often	5	4.46
Rather often	22	19.64
Hard to say	3	2.68
Rather rarely	50	44.64
Very rarely	32	28.57
SUM	112	100
Part B		
Do you use any IT tools to combine data of a different nature or belonging to extensive data sets in such a way as to be able to draw conclusions resulting from the detection of previously unnoticed relationships?		
Definitely not	10	8.93
Probably not	29	25.89
Hard to say	1	0.89
Rather yes	43	38.39
Definitely yes	29	25.89
SUM	112	100
Part C		
How often do you encounter organizational barriers in accessing the data needed to prepare a report, as the data is being held by other organizational units?		
Very often	5	4.46
Rather often	14	12.50
Hard to say	0	0
Rather rarely	50	44.64
Very rarely	43	38.39
SUM	112	100
Part D		
How many changes are needed in the sphere of ways to get information (1 indicates completely unnecessary changes and 5 indicates absolutely necessary changes)?		
1	16	14.29
2	34	30.36
3	35	31.25
4	20	17.86
5	7	6.25
SUM	112	100
Part E		
How many changes are needed in the sphere of data and information processing processes (1 indicates completely unnecessary changes and 5 indicates absolutely necessary changes)?		
1	21	18.75
2	37	33.04
3	23	20.54
4	25	22.32
5	6	5.36
SUM	112	100

Source: own research

in their assessment by choosing the option located in the middle of the scale; 30.36% considered the need for minor adjustments, and 14.29% considered them completely unnecessary.

The fifth question (Table 5.1, Part E) concerned the problem of assessing the need for change in the sphere of data and information processing. In all, 5.36% of the respondents considered the changes to be absolutely necessary, and another 22.32% assessed the situation only slightly better. The indications of the option located in the middle of the scale concerned 20.54% of the cases. The largest group of respondents (33.04%) expressed the opinion that only small changes were necessary, and 18.75% did not see such a need at all.

An analysis of the correlations table (Table 5.2)—correlations that potentially occur between the answers to the five questions asked in the conducted research—shows that, out of ten possible cases of interrelationship, four statistically significant results can in fact be indicated. All cases relate to the relationships between the answers to four questions: Q1, Q2, Q4, and Q5. The answers to Q4 and Q5 are the most strongly correlated. The second in terms of strength is the correlation combining the answers to Questions Q2 and Q5. Then, correlating the answers to Questions Q2 and Q4 needs to be noted. The weakest correlation is between Questions Q1 and Q5. Of the four pairs presented, Q5 plays an important role in three cases.

Challenges Related to Improving Strategic Information Analytics System From Corporate Perspective—Conclusions and Recapitulation

The existence of a group of large enterprises with strategic information analytics systems in their structures should be assessed as a positive circumstance. Such a tool in the management system creates real chances for undertaking an effective competitive struggle on global markets at a time when the entire economy is undergoing transformation under the influence of the Fourth Industrial Revolution. However, merely stating that an organization has managed to implement a SIAS is still not enough to regard the situation as completely satisfactory. The present research

Table 5.2 Correlations linking answers to Questions Q1, Q2, Q3, Q4, and Q5

	q1	*q2*	*q3*	*q4*	*q5*
q1					
q2	0.00				
q3	0.18	-0.04			
q4	-0.17	-0.26**	-0.14		
q5	-0.23*	-0.37**	-0.15	0.46**	

Source: own research

has shown that many companies in this group are struggling with significant limitations and problems: the excessive amount of data available in the system is a problem for 24.1% of the respondents; 34.7% of the respondents do not reach the benefits of data integration—thus, they lose access to very valuable knowledge (which is potentially within their reach); 17% of the respondents experience significant difficulties in accessing data that is already held by the organization and registered by one of its other departments; 24.1% of the respondents notice the need to introduce changes in the sphere of information acquisition; and 27.7% of the total number of employees surveyed see the need to introduce changes in the sphere of data processing.

Of the two spheres included in the study, the sphere that requires changes more is the one that is responsible for data processing. At the same time, there is a significant relationship here, which means that the higher the level of technological advancement in the use of integrated data, the more the need for changes in the sphere of information acquisition as well as data processing processes are perceived. It can also be said that the need for changes in one sphere (ways of obtaining information) entails the need to also make changes in the other (data processing processes). Finally, it was also found that becoming aware of the excessive amount of information that requires processing makes it obvious that it is necessary to introduce changes in the sphere of data processing.

Corporations around the world—and especially in countries such as the United States—are intensively looking for new organizational solutions and will carefully listen to hints on how to strengthen their competitive position in the global market during the changes brought by the era of 4.0. As one of the key practical conclusions of the conducted analysis, one can indicate the postulate of the management of a company undertaking an in-depth reflection on the level of technological advancement in the integration and use of integrated data within the organization they manage. The dependencies detected due to the conducted research allow us to expect that the result of such an assessment will immediately show whether there is a parallel need to improve the way an organization functions in the sphere of information-acquisition methods or in the sphere of data processing. However, the first step that an enterprise should take is to start to constantly monitor the amount of data that is potentially obtainable and, at the same time, allows it to gain valuable knowledge. If at some point the organization realizes that it cannot process all of the available and valuable data, then it is the last moment to implement rapid and radical changes in the 4.0 era.

The entire analysis shows the background set by the Fourth Industrial Revolution on the one hand and the internal difficulties that need to be eliminated to effectively respond to the challenges that are just emerging in a dynamically changing environment on the other. The author hopes that the presentation of the most important determinants that currently stimulate changes in the environment will allow enterprises to recognize them accurately, understand their essence, and take effective action to respond to them in an appropriate manner. The scale and nature of the challenges faced by large corporations competing in the global market mean that, without implementing appropriate system solutions, it will be extremely difficult for them to continue to develop and build a strong competitive position. The presented considerations touch only selected issues that can be pointed out in the area of the discussed subject. However, they are a valuable inspiration and a starting point for further research and discussion among representatives of the scientific community as well as experts dealing with the practical side of doing business.

References

Antonelli, C. (2014). *The economics of innovation, new technologies and structural change*. Routledge.

Becker, J., Knackstedt, R., & Pöppelbuß, J. (2009). Developing maturity models for IT management: A procedure model and its application. *Business & Information Systems Engineering*, *1*(3), 213–222.

Bukowski, L. (2016). System of systems dependability—theoretical models and applications examples. *Reliability Engineering & System Safety*, *151*, 76–92.

Bukowski, L. (2019). Reliable. In *Secure and resilient logistics networks, delivering products in a risky environment*. Springer Nature Switzerland AG.

Chen, Y., & Li, Y. (2018). *Computational intelligence assisted design in industrial revolution 4.0*. Routledge.

Dziekoński, K. (2017). Factors affecting communication quality in project teams. *Przegląd Organiacji*, *3*, TNOiK, 60–66.

Kagermann, H. (2014). Chancen von Industrie 4.0 nutzen. In T. Bauernhans, M. ten Hompel, & B. Vogel-Heuser (Eds.), *Industrie 4.0 in Produktion, Automatisierung und Logistik* (pp. 603–614). Springer.

Kouatli, I. (2016). Global business vulnerabilities in cloud computing services. *International Journal of Trade and Global Markets*, *9*(1), 45–59.

Laloux, F. (2014). *Reinventing organizations: A guide to creating organizations inspired by the next stage of human consciousness*. Nelson Parker.

Lanza, G., Haefner, B., & Kraemer, A. (2015). Optimization of selective assembly and adaptive manufacturing by means of cyber-physical system based matching. *CIRP Annals—Manufacturing Technology*, *64*(1), 399–402.

Larsson, A., & Teigland, R. (2019). *The digital transformation of labor. Automation, the gig economy and welfare*. Routledge.

Manu, A. (2015). *Value creation and the internet of things. How the behavior economy will shape the 4th industrial revolution*. Routledge.

Monostori, L. (2014). Cyber-physical production systems: Roots, expectations and R&D challenges. *Procedia CIRP*, *17*, 9–13.

Rautenstrauch, C., Seelmann-Eggebert, R., & Turowski, K. (2012). *Moving into mass customization: Information systems and management principles*. Springer Science & Business Media.

Ryu, S. L., & Won, J. (2018). The relationship between competitive advantage and the value relevance of accounting information. *International Journal of Trade and Global Markets*, *11*(1/2), 118–126.

Schuh, G., Potente, T., Varandani, R., & Schmitz, T. (2014). Global footprint design based on genetic algorithms—an 'industry 4.0' perspective. *CIRP Annals—Manufacturing Technology*, *63*(1), 433–436.

Schwab, K. (2017). *The fourth industrial revolution*. Crown and Archetype.

Strømmen-Bakhtiar, A., & Vinogradov, E. (2020). *The impact of the sharing economy on business and society. Digital transformation and the rise of platform businesses*. Routledge.

Trstenjak, M., & Cosic, P. (2017). Process planning in industry 4.0 environment. *Procedia Manufacturing*, *11*, 1744–1750.

Tworek, K., Walecka-Jankowska, K., & Zgrzywa-Ziemniak, A. (2019). Towards organisational simplexity—a simple structure in a complex environment. *Engineering Management in Production and Services*, *11*(4), 43–53.

Van Rijmenam, M. (2019). *The organisation of tomorrow: How AI, blockchain and analytics turn your business into a data organization*. Routledge.

Zhou, J. (2013). Digitalization and intelligentization of manufacturing industry. *Advanced Manufacturing*, *1*(1), 1–7.

6 Effect of Integrated IT Systems on Enterprise Competitiveness at Time of "Industry 4.0"

Anna Wolak-Tuzimek and Radosław Luft

Introduction

The efficient information flows in an enterprise are extremely important to management. Without complete, reliable, and rapid information on finance, production, purchasing, and the like, a firm cannot function properly; therefore, appropriate IT systems are necessary. The effective use of IT in enterprise management brings a range of benefits, beginning with those most general like the more efficient management of an entire enterprise and improved information flows to the measurable benefits that improve a firm's performance indicators. The financial standing and operation in other areas are improved. These systems also offer prospects of continuing undisturbed development and the growth of profits. As a result, enterprise competitiveness improves.

By implementing an integrated IT system, an enterprise is able to make key organizational changes. Such actions may streamline the operation of the organization and improve competitiveness.

It is the objective of this chapter to examine the effects of integrated information systems on enterprise competitiveness. Two research problems were also identified: (1) integrated IT systems have a significant impact on enterprise competitiveness and (2) an evaluation of the effects of integrated IT systems on enterprise competitiveness depends on the organizational status and sector of an enterprise.

Integrated IT Management Systems

Each enterprise is an open system of mutual interactions with its external environment; this is characterized by a heterogeneous structure. The management of contemporary enterprises is largely supported with IT management systems that are defined as the computerized parts of an organization's IT subsystem that is responsible for the management support, the receipt and sending of information to and from other subsystems of an organization, and its environment (Szmit, 2003, p. 14). According to Shaqiri (2014, p. 19), IT management systems are designed

DOI: 10.4324/9781003186373-6

to supply information at the appropriate times and to effectively support the decision-making process and other management functions.

The development of informatization and technological progress have become key factors that have driven the development of enterprises in the last decade. Information is crucial to attaining objectives in the short as well as medium and long terms. The rapid use of information from an enterprise's external and internal environments has become the driving force and activator of many a business strategy. Maintenance and persistent improvements in efficiency are key challenges to organizations. Adequate information flows are necessary, and integrated IT systems help enterprises take advantage of the power of information to boost their efficiency (Cherotich, 2017, p. 1).

IT management systems are applications of diverse properties that are essentially IT models of enterprise operation. The systems of production planning and control have evolved since 1970. They have improved and expanded with successive modules, from material and production resource requirements to network-operated resource planning. They have become known as ERP (Enterprise Resource Planning) systems and have effectively replaced their 1990s predecessors. State-of-the-art integrated systems provide for the planning and management of enterprise finances (Lenart, 2005, p. 28). The addition of more modules to MPR II Money Resource Planning standards (e.g., in procurement, production, or sales) has given rise to a new class of systems that deal with organization resource planning (which are referred to as ERP class systems).

Algarni and Alsanad (2018, p. 392) state that an ERP system is a software that manages business processes by allowing an organization to employ a system of integrated business management applications. This is a system for organizing, defining, and standardizing the business processes necessary for effective organization, planning, and control; in effect, an organization can take advantage of its internal knowledge and experience to seek external benefits (Bharadwaj, 2000). In addition, ERP systems enable the accurate modeling of enterprise business processes and resource management in such other aspects as economic or financial condition (Kumar & Hillegersberg, 2000, p. 23). As a consequence, enterprises focusing on continuing their competitive advantage implement and operate ERP systems to streamline their management processes. The maximum extent of integration of all levels of enterprise management, the entire processes of procurement, production, and distribution, and the management of all key enterprise resources by improving information flows and rapidly responding to threats and opportunities are principal objectives of such systems. These are considered the best IT management systems employed by enterprises at present (Esteves & Pastor, 2016).

An ERP system is modular software that consists of independent albeit co-working applications (production, stock, human resources, bookkeeping, business analytics, delivery, sales, engineering, purchasing, and production planning) that integrate processes across all functional areas. Given such a solution and the rapid sharing of information, work is more coordinated and synchronized (Haddara & Hetlevik, 2016; Gool & Seymour, 2019; Haddara, 2018).

ERP systems evolved in parallel with technology development, general integration, and the rollout of WEB 2.0 Internet services. Old ERP systems have some multi-access IT systems to support enterprise management.

The emergence of the ERP II system in 2000 was another stage of development. ERP II is functionally complex and open—a "set of applications specific to a given industry that generate value for customers and shareholders by making available and optimizing processes both inside an enterprise and among its partners." The ERP II system evolved in particular under the influence of new business requirements; these are strongly oriented toward building relationships with customers and business partners. Aside from the ongoing business process management, these systems help to build an organization in a virtual space (Ksielnicki, 2008, p. 315).

The subsequent systems build on the parameters of their predecessors to improve and expand them. Development of IT systems used in enterprise management was as follows (Rzewuski, 2002; Sumner, 2014; Stepanov, 2015; Pietras, 2017; Bytniewski et al., 2018):

1. Reorder point systems (1960s)—Used historical data to forecast future inventory demand; when an item falls below a predetermined level; additional inventory is ordered, designed to manage the high-volume production of a few products, with constant demand.
2. Material Requirement Planning (MRP) (1970s)—Offered a demand-based approach for planning the manufacture of products and ordering inventory; emphasis on greater production integration and planning.
3. Manufacturing Resource Planning (MRP II) (1980s)—Added capacity planning; could schedule and monitor the execution of production plans; manufacturing strategy focused on process control, reduced overhead costs, and detailed cost reporting.
4. Enterprise Resource Planning (ERP) (1990s)—Integrated manufacturing with supply-chain processes across a firm; designed to integrate a firm's business processes to create a seamless information flow from suppliers through manufacturing to distribution; integrates supplier, manufacturing, and customer data throughout the supply chain.
5. Enterprise Resource Planning II (ERP II) (2000s)—Planning of enterprise resources, including expanded functionality and using Internet technologies; integrates supplier, manufacturing, and customer data throughout the supply chain.

6. Enterprise Resource Planning III (ERP III) (2010s)—Planning of enterprise resources, including expanded functionality and using Internet technologies (in particular, mobile devices); building relationships with customers and business partners in virtual space using mobile devices.
7. Enterprise Resource Planning IV (ERP IV) (2020s)—Planning of enterprise resources, including expanded functionality and using artificial intelligence (cognitive software), the Internet of Things (*IoT*), big data, big management, Industry 4.0, and fog computing; building relationships with customers and business partners in virtual space using new IT technologies.

ERP III (third-generation ERP), which emerged in 2010, is another stage in the evolution of integrated systems. It expands the existing ERP II system by improving cooperation with customers by including them in an enterprise's IT system in order to provide them with a direct and active part in the business processes. Application of the Internet (in particular, mobile technologies) has a key role to play in the concept of ERP III. An ERP III system is a model solution that is expected to assure enterprises that an enterprise functions in line with the virtual paradigm (Wang & Clegg, 2010, p. 193).

The ERP III system actively supports direct contact with a firm by making mobile service tools, social media, and other tools available serving on-line communication, both inside and outside of a firm.

ERP III can also be seen as a starting point for the creation and support of the "borderless enterprise" that functions by means of communication support tools such as social media, Internet technologies, and SOA (Fotache & Hurbean, 2014, p. 267).

Signs of an emergent and newly forming ERP IV system are appearing at present that help to plan enterprise resources, including an expanded functionality and using artificial intelligence (cognitive software), the Internet of Things (*IoT*), big data, big management, Industry 4.0, and fog computing. Basic information about the new IT technologies is part of Table 6.1.

Bytniewski et al. (2018, p. 53) believe that ERP IV helps to realize the mechanisms of business globalization and integration among market partners, cooperating organization (banks, national insurance authorities, and broadly defined public administration), and social portals to a greater extent. The capabilities of such a system are aligned with the requirements of the real-time enterprise (RTE), merely a postulate in the earlier versions of these systems.

ERP IV serves to connect the worlds of production machinery, the virtual world, and information technology. Humans, machines, and IT systems automatically share information as production is in progress.

Table 6.1 Characteristics of new IT technologies

Internet technologies	*Notions and characteristics*
Cloud computing	• The model provides a universal and convenient web access on request to a shared pool of configurable ICT resources (e.g., servers, mass storage, applications, platforms, networks), their rapid acquisition and issue involving a minimum effort and interaction with the model supplier; • Computing services offered by third-party providers and available on request at any time, dynamically scalable in response to variable user requirements.
Fog computing	• A new support paradigm for data transmission and processing to support dispersed devices as part of the Internet of Things concept; • Combines the Internet of Things and the computing cloud; • A virtual platform that provides computing capabilities, mass storage, and network services between end devices and a traditional data center of the computing cloud.
Internet of Things	• All and any smart objects capable of responding to their environment, storage and processing of digital information, and transmitting it to other objects (and their users) by means of Internet protocols; • Communication among machines and their autonomous operation based on data they share.
Big data	• Data or data sets which are so large and complex that traditional data processing applications are insufficient for analysis of these data; • Characteristics of the big data: • Volume—gathering of huge quantities of data to the order of terabytes and petabytes; • Velocity—data come in rapid streams that, in connection with business processes, require additional computing capacities for real-time analysis; • Variety—data come from diverse sources and in different formats and are often recorded by means of various models and expressed in any forms, e.g., as numbers, texts, images, sound, and generated in diverse ways; • Veracity—data analyzed and explored contain true information; • Value—essential and valuable data are analyzed, while worthless data are ignored.
Big management	• Realization of the management process addressing new paradigms in conjunction with the big data concept.

Source: authors' own compilation based on Badger et al. (2012), Marr (2015), Billewicz (2016), Hernes and Bytniewski (2017), and Bytniewski et al. (2018)

This means enterprises implementing ERP IV have entered the fourth stage of the industrial revolution.

"Industry 4.0" solutions support personnel by providing access to each and every piece of elementary information at any time and from any place. These actions are made possible by the application of new information technologies that are used as part of the ERP IV system.

The application of the computing cloud offers access to data from various devices and equipment from any location. Any data can also be freely shared among authorized individuals, which considerably improves collaboration and communication (both within and outside an enterprise).

"Industry 4.0" denotes a digital revolution, with data as its most essential part. This is why enterprises take advantage of the new big data technology, which revolutionizes the approach to enterprise management and supplies evidence of a well-designed database; its professional analysis can substantially improve the operations of an enterprise.

Big data is a set of data that is characterized by high volume, variety, real-time streaming, variability, and complexity while requiring the application of innovative technologies, tools, and IT methods to explore them for new and useful knowledge.

An ERP IV-class integrated IT system supports production enterprises that function as part of "Industry 4.0." As part of modular smart factories, cyber-physical systems monitor physical processes, make virtual copies of the physical world, and make decentralized decisions. Such systems communicate, gather data, and cooperate with one another and with humans in real time by means of the "Internet of Things."

Description of Sample

A survey was conducted in September 2019 covering large enterprises that were active in the Mazovian region. As at the end of 2018, the general population consisted of 1,045 entities (GUS, 2019, p. 58). Three hundred enterprises were drawn out of that population in such a way as to guarantee each entity an equal chance of being in the sample. Such a sub-group represents (and is representative of) the entire population; that is, it allows for conclusions regarding the general set. A computer-assisted telephone interview (CATI) produced 168 survey questionnaires that were correctly filled out. Assuming a confidence level of α=95% and a maximum error of β=7%, the results of the analysis are representative of the general population.

The empirical study utilized an original survey questionnaire that consisted of two parts: particulars and data. Three objective (or close to objective) criteria were adopted in the former to characterize the sample. Two issues were determined in the latter: the effects of IT systems in their place on competitiveness and an evaluation of the benefits from the implementation of IT systems. The results concerning the first problem will be presented here.

The effects of IT systems on competitiveness were measured from four perspectives of the Strategic Scorecard; that is, of customer, development, internal processes, and finances defined by means of 16 observable variables (market share, customer retention, customer acquisition, customer satisfaction, customer profitability, staff turnover, staff productivity, staff skills, integrated IT system, staff commitment to the realization of strategy, innovative processes, operating processes, after-sales service processes, revenue growth and structure, cost reduction, and utilization of assets). It is the authors' opinion that these variables are the factors influencing the levels of enterprise competitiveness. The respondents were to determine the impact of IT systems on the particular observable variables on a scale of 1 to 10, where "1" stands for a low impact and "10" represents a high impact.

The questionnaire contained three questions (independent variables) meant to characterize the enterprises. The first addressed its organizational status. A limited liability company was the prevalent form of organization, representing more than 64% of all of the firms. This was followed by joint-stock companies (more than 23%). The sector of the enterprise was the second criterion formally discriminating the sample. Trade and service businesses constituted the largest grouping (67 entities, or ca. 40% of all of the firms), followed by production enterprises (approximately 27%). Enterprises in the financial sector accounted for 19% of all of the entities studied. The territorial scope of the operation was the next particular. The data indicated more than a half (58.92%) were active in both the domestic and international markets.

Methods

The significance of the integrated IT systems (observable variables) was measured and evaluated by means of descriptive statistics. One position measure (arithmetic mean) that characterizes a statistical set regardless of any differences between its units was employed along with a volatility measure (standard deviation) that characterizes a statistical set considering the differences between its units. The results have been tabulated.

Non-parametric methods were applied to the variables measured along variable scales (Gaca, 2016, p. 32). Therefore, Kruskal–Wallis and Mann–Whitney *U* testing was used to determine the effects of integrated IT systems on enterprise competitiveness.

The Kruskal–Wallis test was applied to independent variables coding for more than two, and the Mann–Whitney *U* test was applied to those coding for two.

A zero hypothesis was posited assuming the equality of the average ranks for the particular groups, whereas the alternative hypothesis assumes that the means are different. Based on the test statistics, *p* (the level of significance) was determined and compared to the level of significance α:

if $p \leq \alpha \Rightarrow$, H0 should be rejected and H1 accepted;
if $p > \alpha \Rightarrow$, there are no grounds for rejecting H0.

The acceptance of H0 implies that the levels of a factor have no significant effect on the observed results. A rejection of H0 means that the levels of a factor have a significant effect on the observed results. A factor then discriminates the results; a level of significance of $\alpha = 0.05$ was assumed. The observed p was analyzed using the results compiled with the aid of *Statistica 12*. Its value represents the diminishing reliability of the results. This value helps us assess the likelihood of a given result assuming that H0 is true. p should be greater than set α.

Results

The impact of integrated IT systems on enterprise competitiveness due to independent variables (legal form of enterprise, enterprise sector, scope of firm activities) was presented in Table 6.2. It presents the results of research in the field of descriptive statistics for the four perspectives of the Strategic Scorecard.

The impact of integrated IT systems on enterprise competitiveness was determined to be maximum in joint-stock and limited liability companies. The mean values of the customer perspective in this group of enterprises were 8.20 and 7.24, respectively. These amounted to 7.08 and 6.55 (respectively) for the financial perspective. As far as the development perspective is concerned, the average assessments for these groupings were 8.08 and 7.21, respectively. With regard to the internal process perspective, the respective evaluations averaged 7.38 and 5.82.

On the other hand, minimum scores were assigned to the other unspecified legal forms. They were 4.81 in the case of customer perspective, 3.99 for the financial perspective, and 4.51 and 3.36 in respect to the development and internal process perspectives, respectively.

As far as customer perspective is concerned, maximum volatility was noted in the limited partnerships when measuring the impact of IT technologies for the various enterprise formats. Their standard deviation reached 2.22. The greatest dispersion around the average was recorded for enterprise organizations not specified in the survey regarding the financial and development perspectives (1.74 and 1.94, respectively). On the other hand, the responses varied the least for the internal process perspective in joint-stock companies (0.25).

It was then determined whether these fluctuations could be generalized to the population of large enterprises active in the Mazovian region.

Table 6.2 Mean values and standard deviations for four perspectives on effects of integrated IT systems on enterprise competitiveness with regard to independent variables

Variable	*Customer perspective*		*Financial perspective*		*Development perspective*		*Internal process perspective*	
	I	*II*	*I*	*II*	*I*	*II*	*I*	*II*
Legal form of enterprise								
Joint-stock co.	8.20	0.82	7.08	0.50	8.08	0.42	7.38	0.25
Ltd liability co.	7.24	1.23	6.55	1.48	7.21	1.51	5.82	1.37
Gen. partnership	5.52	1.64	5.07	1.63	5.16	1.89	4.09	1.91
Ltd partnership	6.96	2.22	6.00	1.39	7.20	1.88	6.50	1.46
Others	4.81	1.99	3.99	1.74	4.51	1.97	3.36	1.80
Enterprise sector								
Production	6.58	1.53	6.21	1.44	6.44	1.85	5.52	1.69
Trade and services	6.01	1.76	5.07	1.61	5.64	1.88	4.19	2.05
Financial sector	5.71	1.86	5.48	1.91	5.57	1.95	4.95	1.91
Others	4.86	1.76	4.26	1.60	4.46	2.11	3.33	1.75
Scope of firm activities								
Domestic market	6.03	1.86	5.45	1.75	5.82	1.99	4.72	2.07
Domestic and international market	5.26	1.79	4.67	1.70	4.90	1.97	3.80	1.87

I—Mean value
II—Standard deviation
Source: authors' own compilation

Since the independent variable had more than two codes (five forms of legal organization), the Kruskal–Wallis test was utilized.

Two hypotheses were postulated as follows:

H0: The distribution of the effects of integrated IT systems on the individual perspectives (customer, financial, development, internal processes) is identical for the variable category of an enterprise's legal form.

H1: The distribution of the effects of integrated IT systems on the individual perspectives (customer, financial, development, internal processes) is not identical for the variable category of an enterprise's legal form.

The test results concerning the effects of integrated IT systems on enterprise competitiveness in respect to the organization are listed in Table 6.3.

The conditions determined for all of the perspectives indicate the zero hypothesis needs to be rejected, as the boundary probabilities are lower than the assumed level of significance (α=0.05). All of the found variations are statistically significant; therefore, the volatility for all four perspectives can be generalized to the large enterprises of the Mazovian region. This means the legal organization of enterprises discriminates the impact of integrated IT systems on enterprise competitiveness.

The effects of integrated IT systems on enterprise competitiveness in respect to sectors needed to be analyzed next. Production firms exhibited maximum values for the four perspectives: customer (6.53), financial (6.21), development (6.44), and internal processes (1.75). In addition, the responses of this group displayed a minimum dispersion around the arithmetic mean, as evidenced by the standard deviations for these perspectives (1.53, 1.44, 1.85, and 1.69, respectively).

On the other hand, the lowest values were recorded for those enterprises indicating "other" sectors of their operations. These values ranged within 3.33–4.86 with regard to the four perspectives. Their responses were the most dispersed concerning the development perspective (2.11).

The significance of the effects of integrated IT systems on enterprise competitiveness with regard to the sector was determined by means of Kruskal–Wallis testing.

Two hypotheses were advanced as follows:

H0: The distribution of the effects of integrated IT systems across the perspectives (customer, financial, development, internal processes) is identical for the variable category of the enterprise sector.

H1: The distribution of the effects of integrated IT systems across the perspectives (customer, financial, development, internal processes) is not identical for the variable category of the enterprise sector.

Table 6.3 Results of Kruskal–Wallis test for the impact of integrated IT systems on enterprise competitiveness as far as legal form is concerned

No.	*Zero hypothesis*	*Test*	*Significance*	*Decision*
1	Distribution of effects of integrated IT systems on customer perspective is identical for the variable category of enterprise's legal form	Kruskal–Wallis test	0.000	Reject zero hypothesis
2	Distribution of effects of integrated IT systems on financial perspective is identical for the variable category of enterprise's legal form		0.000	Reject zero hypothesis
3	Distribution of effects of integrated IT systems on development perspective is identical for the variable category of enterprise's legal form		0.000	Reject zero hypothesis
4	Distribution of effects of integrated IT systems on internal process perspective is identical for the variable category of enterprise's legal form		0.000	Reject zero hypothesis

Source: authors' own compilation

The boundary probabilities are less than the assumed level of significance (α=0.05), which means the zero hypotheses must be rejected (Table 6.4). All of the fluctuations detected for the four perspectives are, therefore, statistically significant and can be generalized to the population of large enterprises in the Mazovian region. This implies that the enterprise sector discriminates the impact of integrated IT systems on enterprise competitiveness.

The analysis of the impact of integrated IT systems on enterprise competitiveness as far as the scope of enterprise activities is concerned suggests that firms that are active in the domestic market only attributed higher values to the four perspectives (6.03, 5.45, 5.82, and 4.72, respectively). The standard deviation was high; however, this is a proof that the assigned values were not clustered around the average.

The analysis continued to determine whether the above variations can be generalized to the population of large enterprises in the Mazovian region. The independent variable had two codes (the domestic and

Table 6.4 Results of Kruskal–Wallis test concerning effects of integrated IT systems on competitiveness regarding enterprise sector

No.	*Zero hypothesis*	*Test*	*Significance*	*Decision*
1	Distribution of impact of integrated IT systems on customer perspective is identical for a variable of the enterprise sector	Kruskal–Wallis test	0.000	Reject zero hypothesis
2	Distribution of impact of integrated IT systems on financial perspective is identical for a variable of the enterprise sector		0.000	Reject zero hypothesis
3	Distribution of impact of integrated IT systems on development perspective is identical for a variable of the enterprise sector		0.000	Reject zero hypothesis
4	Distribution of impact of integrated IT systems on internal process perspective is identical for a variable of the enterprise sector		0.000	Reject zero hypothesis

Source: authors' own compilation

international markets); therefore, the Mann—Whitney *U* test was applied (Table 6.5).

Two hypotheses were posited as follows:

H0: The distribution of the effects of integrated IT systems across the perspectives (customer, financial, development, internal processes) is identical for the variable category of the scope of enterprise activities.

H1: The distribution of the effects of integrated IT systems across the perspectives (customer, financial, development, internal processes) is not identical for the variable category of the scope of enterprise activities.

The boundary probabilities are greater than the assumed level of significance (α=0.05), which means the zero hypotheses must be accepted and all the fluctuations detected are not statistically. This means the scope of enterprise operations does not discriminate the impact of integrated IT systems on enterprise competitiveness.

Table 6.5 Results of Mann–Whitney *U* test for the impact of IT technologies on enterprise competitiveness depending on extents of enterprise operations

No.	*Zero hypothesis*	*Test*	*Significance*	*Decision*
1	Distribution of impact of integrated IT systems on customer perspective is identical for a variable of enterprise scope of operations	Mann–Whitney *U* test	0.157	Accept zero hypothesis
2	Distribution of impact of integrated IT systems on financial perspective is identical for a variable of enterprise scope of operations		0.139	Accept zero hypothesis
3	Distribution of impact of integrated IT systems on development perspective is identical for a variable of enterprise scope of operations		0.084	Accept zero hypothesis
4	Distribution of impact of integrated IT systems on internal process perspective is identical for a variable of enterprise scope of operations		0.463	Accept zero hypothesis

Source: authors' own compilation

Recommendations for Entrepreneurs

At the time of Industry 4.0, an enterprise is unable to function properly without the support of IT solutions. Businesses should therefore implement advanced IT technologies like networks (Internet, intranet, extranet), computing clouds, the Internet of Things, big data, or big management. The application of these technologies will help enterprises make decisions in nearly real time based on the enormous quantities of current and valuable data, which offers genuine potential for supporting even the most complicated business processes.

As was shown in the research, integrated IT systems have an impact on the competitiveness of enterprises. The ERP IV system enables enterprise resource planning by using modern information technologies. The use of cloud computing means that the collected data is available at any time, from anywhere in the world, and from any device. Big data enables the processing of large amounts of varied and complex data from various information sources. The use of the Internet of Things gives one the opportunity to increase productivity and improve profitability and cost efficiency through the better use of resources, reduced downtime, and better use of information in decision-making. The effect of these activities is to develop and then maintain a competitive advantage in the market.

The ERP IV system is the basic element of digital transformation and Industry 4.0. Therefore, entrepreneurs should cooperate with specialists in the field of software engineering and implement an integrated information system in their companies using modern information technology.

Conclusion

The economy in a dynamically changing environment is characterized by high competitiveness and rapid transformations. Therefore, enterprises wishing to gain a competitive advantage in the market must be capable of fast adjustment and rapid response. The current standard of IT technologies forces enterprises to seek new solutions in a number of areas.

The dramatic development of information technology has paved the way for effective IT systems based on computer techniques of data collection, processing, transfer, and presentation (i.e., the so-called integrated IT systems). They serve to simulate and analyze a variety of actions and, consequently, improve process planning and management in a number of enterprise areas. The implementation of state-of-the-art IT tools to enterprises provides foundations for enterprise management and is the main source of information aiding managers with decisions that are key to enterprises; this may lead to enhancing enterprise competitiveness.

Enterprises are currently implementing the integrated ERP IV IT system using new IT technologies, for example, computing clouds or big data. The application of big data offers new development prospects to enterprises. Major possibilities include gaining a competitive advantage by creating tailor-made production models that allow an enterprise to offer customers tailor-made products and services and by the optimization of enterprise management (which helps boost profits).

ERP IV expands the extent of automation and even robotizes information processes; it can transform information into knowledge, which improves the efficiency of decision-making processes and, thus, enhances the effectiveness of an enterprise's operations. These actions become part of the "Industry 4.0" concept, understood as a smart combination of a number of IT technologies employed by enterprises.

The following conclusions were drawn from a study of 168 large enterprises in the Mazovian region:

1. Enterprises operated as joint-stock and limited liability companies (88% of all firms examined) attributed maximum scores (on a scale of 1 to 10) to the effects of integrated IT systems on enterprise competitiveness.

The arithmetic mean was within a range of 5.82–8.20. The high values (above 5) of four perspectives were also assigned by production enterprises and those operating in domestic companies.

2. The application of the Kruskal–Wallis test has demonstrated the distribution of the impact of integrated IT systems on the particular perspectives (customer, financial, development, internal processes) is not identical for the variable categories of enterprise legal forms and sectors of operation. The boundary probabilities are lower than the assumed level of significance (α=0.05); therefore, all of the detected fluctuations are statistically significant, and the volatility for all four perspectives can be generalized to the population of large enterprises in the Mazovian region. This means the organization and sector of an enterprise discriminate the effects of integrated IT systems on enterprise competitiveness.

Taking all of these into consideration, it can be stated that Industry 4.0 is a concept that changes the way in which companies operate. This is a combination of three elements: the real world of production machines, the virtual world, and information technologies (e.g., cloud computing, the Internet of Things, big data). An enterprise can gain a competitive advantage by using advanced data analysis and its application in the management process.

A confirmation of this association is the result of the study, which shows that integrated information systems significantly affect the competitiveness of enterprises.

References

Algarni, M., & Alsanad, A. (2018). *Cloud computing and ERP: An academic literature review (2010–2015), social transformation*. Proceedings of the 23rd UK Academy for Information Systems (UKAIS) International Conference St. Anne's College Oxford, pp. 391–405.

Badger, L., Grance, T., Patt-Corner, R., & Voas, J. (2012). *Cloud computing synopsis and recommendations* Special Publication (NIST SP), National Institute of Standards and Technology, Gaithersburg, MD. https://nvlpubs.nist.gov/nistpubs/Legacy/SP/nistspecialpublication800-146.pdf

Bharadwaj, A. S. (2000). A resource-based perspective on information technology capability and firm performance: An empirical investigation. *Management Information Systems Quarterly*, *24*(1), 169–197.

Billewicz, K. (2016). Possibility of internet of things technology implementation in smart power grids. *Energetyka*, *5*, 264–270.

Bytniewski, A., Matouk, K., & Hernes, M. (2018). Ku systemowi klasy ERP IV. *Informatyka Ekonomiczna*, *1*(47), 43–57.

Cherotich, L. M. (2017). *Effect of Imis (integrated management information system) strategy on the achievement of corporate objectives: A case study of the communication authority of Kenya (Ca)*. United States International University-Africa.

Esteves, J., & Pastor, J. A. (2016). Towards a unified ERP implementation critical success factors model. *Proceedings of the Portuguese Association for Information Systems Conference*, *1*(1), 1–9.

Fotache, D., & Hurbean, L. (2014). *ERP III: The promise of a new generation*. The Proceedings of the IE2014 International Conference, Bucharest, pp. 265–271.

Gaca, R. (2016). Terminy i zagadnienia statystyczne i ekonometryczne w wycenie nieruchomości. *Problemy rynku nieruchomości*, *1*(45), 31–35.

Gool, S., & Seymour, L. (2019). *Managing enterprise resource planning system customisation post-implementation*. Proceedings of the 20th International Conference on Enterprise Information Systems, SCITEPRESS—Science and Technology Publications, Lda, pp. 111–119.

GUS. (2019). *Zmiany strukturalne grup podmiotów gospodarki narodowej w rejestrze regon 2018 rok*. Warszawa.

Haddara, M. (2018). ERP systems selection in multinational enterprises: A practical guide. *International Journal of Information Systems and Project Management*, *6*(1), 43–57.

Haddara, M., & Hetlevik, T. (2016). Investigating the effectiveness of traditional support structures & self-organizing entities within the ERP shakedown phase. *Procedia Computer Science*, *100*, 507–516.

Hernes, M., & Bytniewski, A. (2017). Towards big management. In D. Król, N. Nguyen, & K. Shirai (Eds.), *Advanced topics in intelligent information and database systems, ACIIDS 2017* (Vol. 710, pp. 197–209). Studies in Computational Intelligence.

Ksielnicki, J. (2008). *MIS systemy informatyczne zarządzania*. Wydawnictwo Placet.

Kumar, K., & Hillegersberg, J. (2000). ERP expertiences and evolution. *Communications of the ACM*, *4*, 23–26.

Lenart, A. (2005). *Zintegrowane systemy informatyczne klasy ERP: Teoria i praktyka na przykładzie systemu BAAN IV*. Wydawnictwo Uniwersytetu Gdańskiego.

Marr, B. (2015). *Big data: The 5 Vs everyone must know*. www.linkedin.com/pulse/20140306073407-64875646-big-data-the-5-vs-everyone-must-know.

Pietras, P. (2017). Istota zintegrowanego systemu zarządzania klasy ERP. *Logistyka*, *6*, 1497–1501.

Rzewuski, M. (2002). Ewolucja systemów zarządzania—ERP II—Nowy stary gatunek. *PC Kurier*, *20*.

Shaqiri, A. B. (2014). Management information system and decision-making. *Academic Journal of Interdisciplinary Studies MCSER Publishing, Rome-Italy*, *3*(2), 19–23.

Stepanov, D. Y. (2015). ERP-system implementation as a way to improve business processes. *Fundamental Problems of Radioengineering and Device Construction*, *15*, 156–158.

Sumner, M. (2014). *Enterprise resource planning*. Pearson.

Szmit, M. (2003). *Informatyka w zarządzaniu*. Warszawa.

Wang, Y., & Clegg, B. E. (2010). Enterprise management and ERP development: Case study of zoomlion using the dynamic enterprise reference grid. In J. E. Quintela Varajão, M. M. Cruz-Cunha, G. D. Putnik, & A. Trigo (Eds.), *Enterprise information systems, centerish. Communications in computer and information science* (Vol. 109, pp. 191–198). Springer.

7 EFQM RADAR-Based Assessment of RFID System as Part of Industry 4.0 Implementation—A Case Study of a Production Plant

Joanna Martusewicz, Karol Szewczyk and Arkadiusz Wierzbic

Digital Transformation Around the World and in Poland

Digital transformation is a tool of Industry 4.0 (the Fourth Industrial Revolution) as well as a vision of the future for most enterprises. It is a concept for connecting the world of production machines with the world of the Internet, information technology, and people. The flow of information among all of these elements should be automated and connected within the digital ecosystem.

Companies of all sizes can take advantage of innovative digital solutions and transform their enterprises into "smart factories"—the intelligent factories of the future. Processes are carried out automatically, production is optimized, and all areas of activity intertwine smoothly and work together to ensure the maximum profit for a company. According to Machado et al. (2020), the key to this approach is to understand the network of connections that must exist in an organization between new technologies, products, the value chain, and the business model of the organization. According to the PwC report (PWC, 2020), a key role is played by a broader view of the processes within an organization and its structure, that is,

- Transformation of product and service offers;
- Vertical and horizontal integration of value chain;
- Optimization of customer relations;
- Building a new business model.

Poland is at the beginning of this road. Studies on the levels of automation of Polish production plants prove that, for the managers of Polish factories, the challenges of the Third Industrial Revolution related to microelectronic technologies remain largely valid. Only 15% of Polish

DOI: 10.4324/9781003186373-7

factories are fully automated (Zieliński, 2016), and 76% of the respondents declared partial automation. Only a small group of factories (i.e., 6%) use IT systems for operational management and production control (the manufacturing execution system [MES] class).

The next important issue to be discussed is the fact that the following mostly disturbing attitudes have emerged from the management of various sizes of manufacturing companies in Poland over the past 3 years (Gracel, 2018a, p. 18). The first group (approximately 30%) are managers who are not interested in the market environment and focus on current operational tasks. Their prevalent attitude is expressed by the words, "I have not heard. I have no time." The second most numerous group (45%) are managers who have encountered the idea of Industry 4.0 but have too many current operational issues for now. This attitude is expressed in the words, "I heard. I have no time." The third group (about 20%) are managers who are aware and understand the opportunities and threats and work together on the concept with their teams. The statement that fits well here is, "I heard, I understand. We are working on the concept." (Gracel, 2018b, p. 35). The fourth group is well-prepared for the coming of the Fourth Industrial Revolution. On the one hand, there is manager involvement in launching investment programs and the strategic empowerment of the initiative, and on the other hand—team involvement in conceptual and implementation work. In practice, they are branches of international production groups with German, Italian, or Scandinavian capital. Their attitude can be illustrated with the words, "we have a road map. We have started the first projects." This is a small group—about 5% of Polish companies (Gracel, 2018b, p. 39).

The last and by far the least numerous group are managers—leaders who have an idea about redefining the industry or industry category in which they operate. They constitute about 0.1% of the population. This very courageous attitude requires a well-thought-out vision for the industry combined with a real concept of convincing and engaging the existing market participants (clients, partners, and competitors) to change and plan to finance the transformation that will not destroy their current business. This analysis shows that there is no other way of development in today's world.

The EFQM Model as a Framework for Assessing the Implemented Industry 4.0 Solutions

The EFQM model is one of the methods that supports the holistic preparation of an organization for Industry 4.0. The model of excellence helps modern organizations measure, anticipate, and monitor the needs and requirements of the interested parties and tracks the achievements of other organizations. The results achieved are the basis for the implementation of improvement activities in the organization (EFQM, 2013).

The evolution of the path to excellence was shown in the work by Hermel and Ramis-Puyol (2003), where one can find an anticipation of the concept of integrated improvement—the so-called Excellence 5.0. This concept is based on the integrated improvement of an organization based on a systematic view of management, taking into account all of its dimensions (Stańczyk-Hugiet, 2014).

According to Carvalho et al. (2019), the values and goals of the model focus mainly on improving the competitiveness of companies on the domestic and global market as well as emphasizing the importance of flexibility, creativity, and an innovative approach to management. Key areas of improvement are as follows: leadership, employees, clients, strategy orientation, process management, etc. Strong points of the stakeholders, the environment, and the long-term orientation of a company's success are also important aspects. The key element is the result of the organization recognized in the context of customers, employees, and society (but above all, the business results). Socio-cultural conditions are of fundamental importance when creating the principles of excellence models. According to Talwar (2011), it is important to properly include the culture and customs of a given country in the organizational culture being built.

Due to self-assessment, an organization can obtain a picture of where it is and what the basis is for determining the path to its continuous improvement. The proposed self-assessment of enterprises based on models of excellence (compared to other assessment systems) takes a richer range of issues into account that comprehensively relate to various activities for quality and excellence. Criteria regarding "results of operations," which impose the necessity to make an assessment, are of key importance, for example, a company's level of competitiveness on the market as well as its ability to meet the expectations of its shareholders. Non-financial factors such as effective working time and productivity are also taken into account during the assessment (Kacała & Kołaczyk, 2012). The model gives organizations the opportunity to improve their activities in areas such as

- Strategic management—by creating a set of key results necessary for monitoring progress in the implementation of the vision, mission, and strategy, thus enabling leaders to make effective decisions;
- Customer value—by clearly defining, communicating, and engaging it in the product-design process;
- Developing leaders' skills to adapt, respond, and acquire the involvement of all stakeholders to ensure the lasting success of an organization;
- Organizing the processes to enable the implementation of a strategy of balance between the strategic goals of an organization and the expectations and aspirations of its employees;

- Seeing stakeholders as a potential source of creativity and innovation—by developing and engaging them in cooperation;
- Fostering partner relationships by shaping lasting and mutual benefits outside the supply chain;
- Taking responsibility for the activities undertaken by an organization and managing their impact on society.

Improving an organization's activities in these areas should ensure the greater efficiency and more positive results of running the business in the short and long term.

Research conducted by Kaynak (2003) and Wiedenegger (2012) clearly indicates better financial results among organizations using models of excellence as compared to similar non-users. In a meta-analysis carried out in 10 out of 15 studies, a statistical correlation between winning an award and business success was determined. Therefore, it can be assumed that a company that operates based on a model of excellence works more efficiently, effectively, and profitably than other companies in the industry and on the market. The key problem among senior management is the pace of change. Models of excellence are not models of quickly repairing selected elements of an organization but usually relate to long periods of time and involvement in sustainable activities at the system and strategic levels.

The Deloitte report (Deloitte, 2020) suggests that management is aware of the opportunities offered by Industry 4.0 and that enterprises are planning intensive investments in the sphere of digital transformations. At the same time, however, managers are not entirely sure how to take advantage of these opportunities; strategic and operational inconsistencies threaten to put a stop to these efforts. There is a clear lack of consistency in terms of strategy, transformation of the supply chain, preparation of staff, and investment-friendly factors. This shows that companies want to introduce changes in the digital sphere, but that it is difficult for them to find the right path, namely the optimal balance between the benefits of ongoing operations and the opportunities offered by Industry 4.0 in terms of the innovation and transformation of the business model. The report identifies key issues with which managers must deal; these are problems with shaping the strategy, adjusting the supply chain, and building human resources.

In all studies, management indicates that digital transformation is one of the most important strategic goals in their organization. However, even though the respondents seem to understand its significance, they perceive changes in this regard in terms of opportunities to increase profits. This means that, while the respondents can associate operational improvements with strategic development, they do not necessarily associate digital transformation with the increase in revenues resulting from new products and business models based on research and development.

Many seem to see the digital transformation as a "defensive" investment aimed at making a business secure rather than developing it.

As efforts are being made to transform an organization into an integrated enterprise capable of operating in an increasingly digitized world, management has many opportunities to build a more coherent, internally cooperative, flexible, and intelligent business and to find a path that will truly realize the promises of Industry 4.0.

Case study—Presentation of the Implemented Solution

In this part, the authors focus on improving the activities inside the internal logistics with the use of RFID solutions as a basis for introducing Industry 4.0 solutions to a manufacturing plant. A pilot project for e-kanban optimization for a pull system is described later. The researched organization is a Polish plant of an international company with a large product portfolio that is located near Wroclaw. The product scope includes producing safety products. The plant's manufacturing processes cover internal machining as well as sub- and final-assembly operations. For the Wroclaw plant, the RFID project is implemented on the sub-assembly of a certain product named for the purposes of this study product A. The plant's management is interested in developing a standard solution that will be introduced for an e-kanban specific to repetitive manufacturing processes. With the current process, the sub-assembly backflush confirmation on the assembly area is carried out fully manually by a POUP (point-of-use provider) depending on the requirements from the final-assembly line. A planning process is not in place—the production needs are transferred verbally to the respective production line without any stock control. The target situation is to have a consumption control process in place with an RFID e-kanban solution set. Production planning for the sub-assembly line is carried out according to the kanban consumption by the main lines. The kanban status changes to "FULL" (a production confirmation of 10 digits in the SAP system) while transferring the goods to the supermarket. The kanban status changes to "EMPTY" when the goods are taken to the mainline for further assembly. Based on the lot formation quantity, changeovers are scheduled. The physical supermarkets are adjusted to new standards (including steel construction with dedicated lanes and space markings).

Support in this project was given by the headquarters—the RFID center of competence, which recommends various portfolios of preferable devices and gives the necessary consulting before the final purchase. Starting with RFID readers/gates and finishing with single tags/chips enables to support each process all in one environment. The plant decided to use one of the most popular and universal tags in the automotive industry—"RFID UHF SmartracDogBone"—easily available on the local market. In accordance with the manufacturing area layout and specification, it was possible to use the standard designation UHF FEIG Mid-Range Reader. This can be used for all kinds of applications that require a medium read range and good performance (a read range of up to 4 m*). In order to secure the manufacturing processes in case of RFID malfunction (i.e., LAN issue, crosstalk connection), a standard kanban process is put into place via ALPE SCAN. By the standard production kanban functionality in SAP, the consumption control process can be fully ensured. Due to the difficult manufacturing conditions, the Motorola Workabout Pro 4 gun mobile device was chosen. The connection to SAP is established by the Wavelink Industrial Browser.

The adaptation of the whole manufacturing area starting from value stream design as well as the raw and finish goods loops was calculated to define the supermarket space. As a consequence, a layout change was performed for the assembly part supermarket to ensure enough physical space for the required number of kanbans. Additionally, a new SAP process implementation was necessary. There are two production processes defined in order to ensure production confirmation in the sub-assembly area. The implementation of kanban with ALPE SCAN and consumption control meant developing the end user's knowledge about the general principles of the lean production approach, including the physical handling of the kanban cards and use of the new ALPE functionality. As per the new process/solution, PK_kanban was treated as a precondition project before RFID implementation. For learning the kanban status change, an internal process was described for better understanding. While fulfilling a production order, the backflush confirmation of ten-digit kanban status changes to "FULL." If there are no issues with the kanban status change (a backflush confirmation of ten digits in the background), a follow-up message will pop up. After the correct booking, the kanban card needs to be placed on a dedicated trolley with the semi-finished products and transferred to a dedicated storage place in the supermarket. Finally, hardware approved by the center of competence RFID is used for the rollout—including 15 SICK RFU620 fixed devices. Due to the limited size, it was not possible to present more details about the material flow improvement based on pull system principles.

Case study—Assessment of the Implemented Solution Versus RADAR for Enablers of EFQM Model 2013

This part focuses on an analysis of the enablers based on the RADAR approach from the EFQM model and the introduction of the Industry 4.0 solution with the use of RFID technology for improving the internal material flow according to the pull principle. The respondents were managers responsible and involved in this project. This step focuses on the elements, attributes, and guidance for the Approaches, Deployment, Assessment and Refinement, which are detailed as follows:

The element "Approaches" contains two attributes:

- "Sound" with the guidance: "The approaches have a clear rationale, based on the relevant stakeholder needs, and are process based.";
- "Integrated" with the guidance: "The approaches support strategy and are linked to other relevant approaches.".

The element "Deployment" also contains two attributes:

- "Implemented" with the guidance: "The approaches are implemented in relevant areas, in a timely manner.";
- "Structured" with the guidance: "The execution is structured and enables flexibility and organizational agility.".

The element "Assessment and Refinement" contains three attributes:

- "Measurement" with the guidance: "The effectiveness and efficiency of the approaches and their deployment are appropriately measured.";
- "Learning & Creativity" with the guidance: "Learning & creativity is used to generate opportunities for improvement and innovation.";
- "Improvement and Innovation" with the guidance: "Outputs from measurement, learning & creativity are used to evaluate, prioritize and implement improvements and innovations." (EFQM, 2013).

The project was started due to the experience and knowledge of the team from headquarters that supports the shaping and development of new technologies within the group. As early as 2016, it was possible to propose the plant in Wroclaw to the list of locations for which the implementation of new I4.0 solutions supported by RFID technologies was planned. Already in January 2017, a pre-scoping of the pilot project was organized where the entire value stream was analyzed in the context of possible improvements along with a reference to the available solution sets supported by RFID (the best practice examples from the other company divisions). Ready solution sets were available in the CoC (center of competence) RFID department, published on the company intranet as state-of-the-art developments for worldwide use within the group. After an internal team analysis and expert discussion, it was decided to use one of the standard RFID solutions for the production kanban with the pull principle process in the pilot sub-assembly area on the mainline in Value Stream 1.

Based on an analysis of the respondents' answers, the authors assessed that the attributes for the Approaches (sound, integrated) were partially performed.

The implementation of the project was well-described and well-structured. A cross-functional team was working on the topic with six team members, which took 4 months. The project started in May 2017. After obtaining management permission for the implementation of the project activities with the support of the central team (service agreement), verification commenced of the basic processes/hardware settings/software gaps in the Wroclaw plant according to the checklist called "maturity assessment for implementation of RFID technology." The RFID implementation checklist was used prior to implementation activities. This document was created according to the know-how gained during years of experience of the center of competence and central logistic department.

In the document structure, the central team gathered all of the necessary data to verify the status of the local preparations and the implementation of all of the necessary activities on an ongoing basis for a successful "go-live." The final solution was described in the business specification workbook, stored on a shared drive in the project folder, which contains data with details concerning the technical specifications and organizational needs. There is the possibility of making easy and flexible updates.

Based on an analysis of the respondents' answers, the authors assessed that attributes for Deployment (implemented, structured) were fully performed.

During the analyses of the project documentation for the assessment and refinement, the study found positive points for the design phase with documents from the RFID technology and RFID implementation checklist and OPL design. Each step of the project had a specific quality gate assessed by a checklist of whether this is in line with the requirements or not. The knowledge gained with this pilot project was later used during the implementation in other areas. For the purpose of monitoring the correctness of the kanban process with the pull principle, the local visualization was developed.

Part of the talking plant project was the inspiration gained during the first RFID implementation. There is a visualization available on every final-assembly/pre-assembly line, consisting of various graphs used for the kanban process monitoring, logistic performance (transport kanbans), and finished goods stock. This point is a common conclusion in the lean manufacturing assessment and improvement along with the innovations. During the launch of subsequent RFID loops, the authors developed a visualization that monitors any critical situations and shows whether the kanban system is working properly.

Based on an analysis of the respondents' answers, the authors assessed that the attribute for the Assessment and Refinement (Measurement) was partially performed.

While analyzing the learning & creativity, the research showed that the internal documentation was developed as a know-how base for further process improvements and the implementation of the pull principle in the Wroclaw plant based on the experience gained during the implementation of the project in the pilot area. The document is described as an RFID handbook. Due to an in-depth insight into all of the process details and examples from the use cases of the implementation on the shop floor, the document has been widely recognized by the central departments and shared with the other plants in the group.

Based on an analysis of the respondents' answers, the authors assessed that this attribute was fully implemented.

As far as the attribute of Improvement and Innovation is concerned, due to the know-how acquired during the implementation of the first production kanban process with RFID in the area of the sub-assembly presses, subsequent implementations were planned. Implementation in the rest of the plant in the areas of machining and final assembly was finished with internal resources. RFID technology is being also developed with the transport kanban and is the basis for the talking plant visualization project.

Based on an analysis of the respondents' answers, the authors assessed that the attribute of Improvement and Innovation was fully implemented.

Case study—Assessment of the Implemented Solution Versus RADAR for the Results of EFQM Model 2013

This section focused on an analysis of the results based on the RADAR approach from the EFQM model. The analysis was performed during interviews with plant management and the internal logistic manager. Data and indicators for the project concerning the introduction of RFID technology to the material flow according to the pull principle were reviewed. In this step, the authors focused on the elements, attributes, and guidance for the Results (Relevance & Usability, Performance) details found here.

The element "Relevance & Usability" contains three attributes:

- "Scope & Relevance" with the guidance: "A coherent set of results, including key results, are identified that demonstrate the performance of the organization in terms of its strategy, objectives and the needs and expectations of the relevant stakeholders.";
- "Integrity" with the guidance: "Results are timely, reliable & accurate.";
- "Segmentation" with the guidance: "Results are appropriately segmented to provide meaningful insights."

The element "Performance" contains four attributes:

- "Trends" with the guidance: "Positive trends or sustained good performance over at least 3 years.";

- "Targets" with the guidance: "Relevant targets are set and consistently achieved for the key results, in line with the strategic goals.";
- "Comparisons" with the guidance: "Relevant external comparisons are made and are favorable for the key results, in line with the strategic goals.";
- "Confidence" with the guidance: "There is confidence that performance levels will be sustained into the future, based on established cause & effect relationships." (EFQM, 2013).

According to the first element of Relevance and Usability, attribute Scope and Relevance, one can observe that the factory in Wroclaw has found a wide application throughout the entire value stream in the plant due to the experience gained in the field of the usability of RFID technology in the logistics and production processes (pull principle)/material staging. The initial launch in the pilot area (kanban production process with the pull principle) gave rise to all future developments focused on streamlining the material flow from admission to the warehouse (warehouse-production orders) to the final finished product declarations in SAP. Due to the small investment, financial savings were also noticeable. The kanban transport process with RFID (material staging) launched in all production areas in the Wroclaw plant brought measurable benefits in many areas related to a 33% reduction in the time needed to complete a contract. The time needed to order a single component was reduced by about 40% (410,104 orders per year, which results in a saving of 1,140 hours). The inventory on production was reduced by increasing the frequency of component delivery. The minimum and maximum stocks were defined (min/max on components). The losses on annual inventories were reduced. Based on an analysis of the respondents' answers, the authors assessed that this attribute (Scope and Relevance) was fully implemented.

Monitoring the results (the correctness of the process along with the stabilization of the transport kanban process), the logistics performance monitoring system for the entire plant has been expanded. Due to this, the results are available in real time in every area: the production kanban process with RFID. The benefits resulting from the commissioning of the pull system have significantly improved the material flow process between the sub-assembly and the final lines. It optimized the safety stock (min/max) and introduced high inventory transparency. The delivery performance also increased—the details (the logistic KPI tree) were shown during the interview. The work-in-progress stocks were reduced by eliminating the production of unnecessary parts (overstock). Improvement of the OEE indicator was presented in the KPI tree for productivity during the assessment. An additional benefit shown was the improvement in yearly inventory losses (in 2017, it was 590kPLN, 2018–397kPLN, 2019–29kPLN). Reducing the losses on the annual inventories had a significant influence on forecasting and plant results at the end of the year.

Based on an analysis of the respondents' answers, the authors assessed that the attribute Integrity was fully implemented.

Credibility—all of the presented results are reliable and confirmed and reported to the Wroclaw plant management. OEE reliable results are available in the OEE Report which is created in Tableau Software. In this report, one can see a visualization of the plant and assembly lines as well as the results of the OEE indicator as compared to the target (green—lines fulfilling target; red—needs for corrective actions). The next example can be shown with talking plant visualizations. These show the data for each production area: the status of finished goods stock, sub-assembly high runners stocks, production planning data based on the kanbans, and potential gaps. These results are shown on big LCD screens.

Segmentation—due to the large amount and comprehensiveness of the data, the appropriate transparency of the information and indicators was ensured. For this, one can use examples from the OEE tableau report, talking plant visualizations, and the KPI tree. Based on an analysis of the respondents' answers, the authors assessed that this attribute was fully implemented.

In the element of Performance, the attribute of Trends was evaluated. The stability of the processes shows the great commitment of the entire team and the extensive thematic knowledge. Due to this, support departments directly involved in improvements easily implemented subsequent projects for the new commissioning of production lines/areas. After the stabilization phase for all improvement projects devoted to the subject of the pull principle from the production and transport kanbans, a positive trend in the plant's results was almost immediately observed. It is possible to see the details in the OEE tableau (data from 2018 to the present) and BPS assessment (data from 2015 to 2020). Based on an analysis of the respondents' answers, the authors assessed that this attribute (Trends) was fully achieved.

When the measurable targets are taken into account, the company is constantly improving its processes and working on its stability. Regular CIP (continuous improvement process) workshops help to set new trends, thus allowing for constant improvement and the achievement of higher targets. Based on an analysis of the respondents' answers, the authors assessed that this attribute (Targets) was partially implemented.

Comparison—based on an analysis of the respondents' answers, the authors assessed that this attribute was not implemented.

Confidence—due to the strong customer orientation and the need for continuous improvement at the Wroclaw location, the company is sure that they are implementing projects and achieving the assumed results at the highest possible levels. The young and dynamic WwP team is dedicated to the achievement of any assumed/set goal. Based on an analysis of the respondents' answers, the authors assessed that this attribute was fully implemented.

Conclusions

The opportunities that arise from the Industry 4.0 solutions, the case study of the improvement of the internal logistics using RFID technology, and the assessment using the EFQM-based RADAR methodology are the topics highlighted in these conclusions.

As stated by K. Schwab, the leader of the World Economic Forum, "We stand at the threshold of a technological revolution that fundamentally changes the way we live, work, and co-exist. In its scale, scope, and complexity, this transformation will be something that humanity has never experienced before. . . ." It is important to use this opportunity which currently occurs, such as Industry 4.0, can push companies toward the use of the latest technologies and IT solutions, which in the end will improve their financial results and increase their competitiveness. From the research, it is already known that one of the key elements of this transformation is the use of data sets—the data obtained from production processes (namely, industrial automation systems). Data analytics solutions can improve the production process—this determines the final financial result.

The case study of a mature automotive plant employing approximately 1,000 people and being a part of an international manufacturing network of market leaders in safety systems was presented. From this perspective, the company had an opportunity to use the best practices from Industry 4.0 and implement their own solutions. Over the last 4 years, several projects have been started; one is presented here, as the focus was put onto internal logistic improvements using RFID solutions. This example is addressed to the top and middle management from manufacturing and logistic departments. Such or other similar solutions can be used in similar plants as well as in mid-sized companies from the automotive sector as a benchmark or inspiration for a tailored system according to the company's needs. The benefits increase with the implementation of the next Industry 4.0 solutions in areas other than the pilot one. Automatic booking in an SAP system that shows improvement in time and transparency without human error. This is the basis for the continuous improvement of process material flow and data accuracy in an ERP system. The new data-driven solutions can become an opportunity for saving money and creating process improvements.

The introduction of a new approach in plant routines and improvements in management coming from Industry 4.0 solutions should be assessed not only from the perspective of KPIs but also by the use of interdisciplinary metrics and performed assessments. This is an activity that can give a different perspective and show potential for improvements that are not clearly visible from a KPI follow-up and can become a real added value for a company's upcoming projects.

Currently in the market, there are several models that can be used; of these, the most popular are the Baldrige Excellence Framework and the EFQM model. In this case study, EFQM Model, 2013, was used because in the past, the researched Wroclaw plant had experience with similar assessments and took part in regional and national quality awards (based on the EFQM model). The RADAR logic that was used to assess the implementation of an RFID system proved to be a useful management tool. The researched managers understood the idea of the RADAR attributes and could assess the Industry 4.0 solution using this set of reference statements. Although some RADAR attributes were not assessed as fully achieved, the picture that arises from the research proves that a technical solution was implemented in a controlled way, and this created measurable benefits. What is even more important is that the researched managers see the RFID system (a part of the Industry 4.0 initiatives) as a way of securing good results in the future. In the area of enablers, the attributes for assessment and refinement were assessed as partially performed based on an analysis of the respondents' answers. Members of the working team were pleasantly surprised on discovering that the EFQM model is a part of the company's DNA. This statement may lead to the conclusion that, in the future, the company will go further on the road to Industry 4.0 and the digitization of manufacturing and future projects; their implementation will be assessed with the use of the EFQM toolkit.

References

Carvalho, A. M., Sampaio, P., Rebentisch, E., & Saraiv, P. (2019). 35 years of excellence, and perspectives ahead for excellence 4.0. *Total Quality Management & Business Excellence*, *30*, 2–3.

Deloitte. (2020). Raport. *Paradoksy przemysłu 4.0-Wyniki globalnego badania Deloitte*. https://www2.deloitte.com/pl/pl/pages/technology/articles/4-rewolucja-przemyslowa-raport.html

The EFQM Excellence Model. (2013). *Brussels: European foundation of quality management, 2013*. www.efqm.org/index.php/efqm-model/download-your-free-copy/

Gracel, J. (2018a). Postawy Menedżerów wobec przemysłu 4.0, w: W stronę przemysłu 4.0. In D. Zaraziński & P. Szymczak (Eds.), *Praktyczny przewodnik* (pp. 24–38). HBRP.

Gracel, J. (2018b). Cztery archetypy Fabryki przyszłości, w: W stronę przemysłu 4.0. In D. Zaraziński & P. Szymczak (Eds.), *Praktyczny przewodnik* (pp. 38–48). HBRP.

Hermel, P., & Ramis-Puyol, J. (2003). An evolution of excellence some main trends. *The TQM Magazine*, *15*(4), 230–243.

Kacała, J., & Kołaczyk, E. (2012). Efektywność wdrażania modelu doskonałości EFQM. *Prace Naukowe Uniwersytetu Ekonomicznego we Wrocławiu*, *274*, 123–134.

Kaynak, H. (2003). The relationship between total quality management practices and their effects on firm performance. *Journal of Operations Management*, *21*(4), 156–178.

Machado, C. G., Winroth, M. P., & Elias Hans Dener Ribeiro da Silva. (2020). Sustainable manufacturing in industry 4.0: An emerging research agenda. *International Journal of Production Research*, *58*(5), 462–484.

PWC. (2020). Raport, Przemysł 4.0 czyli wyzwania współczesnej produkcji. www.pwc.pl/pl/pdf/przemysl-4-0-raport.pdf

Stańczyk-Hugiet, E. (2014). Ewolucja koncepcji doskonałości jako egzemplifikacja różnicowania. *Prace Naukowe Uniwersytetu Ekonomicznego we Wrocławiu*, *359*, 48–63.

Talwar, B. (2011). Business excellence models and the path ahead. *The TQM Journal*, *23*(1), 21–35.

Wiedenegger, A. (2012). *Unternehmensqualitat—Was ist das? Eine theoretische und empirische Untersuchungwelchen Anforderungen Unternehmenentsprechensollten* (PhD dissertation). WU Vienna University of Economics and Business, Vienna.

Zieliński, M. (2016). *Przemysł 4.0 w Polskich Fabrykach. Raport*. www.astor.com.pl/biznes-i-produkcja/raport-przemysl-4-0-polskich-fabrykach/

8 Internet of Things as One of Industry 4.0's Trends—Chosen Theoretical and Empirical Aspects

Anna Barwińska-Małajowicz and Patrycja Żegleń

Introduction

The transition from Industry 3.0 to Industry 4.0 demands many comprehensive analyses that enable the understanding of nonreversible changes. There are some elements of these changes that also cause various effects. One of these is the Internet of Things (IoT), which enables the communication between machines (M2M). This facility generates a production environment that becomes more and more human-independent. The main aim of this chapter is to present the essence of one of the most important areas of Industry 4.0 (i.e., the Internet of Things), both in theoretical and practical aspects. The theoretical aspects are presented on the basis of studies of the literature of the subject. The practical aspect is presented based on companies in Europe whose activity is connected with Industry 4.0. The following main research methods were applied in the chapter: the desk research method and own research analysis, where in-depth interviews (IDI) were conducted on the basis of a non-standardized interview questionnaire. The interview was conducted with the representative of a company whose activity enters into the Industry 4.0 concept.

Internet of Things—Theoretical Aspects on Basis of Literature on Subject

The idea of the Internet of Things has met with great interest from governments, scientific technology enterprises, and scientific research institutions (Guo et al., 2014; Weber, 2010; Sarma et al., 2000). The literature on the subject gives many definitions, examples, and discussions about what the Internet of Things is. The leading idea of the term is that it is going to revolutionize our lives in all of the sectors we operate, starting from transport to health (Maple, 2017, p. 155).

The Internet of Things is one of the areas connected with Industry 4.0 (Xu & Duan, 2018, p. 148) that originates from a project for advanced manufacturing vision supported by the German government in 2011

DOI: 10.4324/9781003186373-8

(Lasi et al., 2014; Xu et al., 2018). Since then, it has become a widely used concept in many spheres of our lives. Industry 4.0 means a new reality of the contemporary economy, as digital progress, transformations, and widening networks are challenges for many companies where one can observe long-range effects such as networks of machines, organizational structural changes, integration and employee qualifications in network processes, and using IT solutions toward companies' valuable data protection. Industry 4.0 significantly changes products and production systems that concern planning, processes, operations, and services; it permeates global value networks, resulting in the management and organization forms of future workplaces creating new business models. Industry 4.0 (also defined as the Fourth Industrial Revolution) is used in many European countries. In other countries (mainly in English-speaking ones), the term is known as the Industrial Internet, the Internet of Things, or smart factory. Except for the non-uniform name, the characteristics of Industry 4.0 are not the same in all countries; it depends on the interpretation and perception by various environments (www2.deloitte.com).

Studies of the literature on the subject make it possible to create the following areas of Industry 4.0 trends:

1. Cloud computing;
2. Big data;
3. Internet of Things;
4. Digitization, simulation;
5. Augmented reality;
6. Human-robot cooperation;
7. Wide palette of sensors and its integration with artificial intelligence (Sąsiadek & Basl, 2018, pp. 189–190).

The driving force of today's changes, relatively commonly used are the first two solutions mentioned above (cloud computing and big data). Nevertheless, the fundamental issue of Industry 4.0 is the Internet of Things, an omnipresent human, thing, and machine network that should bring many new services and offers by definition. The virtual world seamlessly meets real-world objects (Ashton, 2010, www.rfid journal.com).

The "Internet of Things" term became popular in order to emphasize global infrastructure Visio that joins objects/physical items with the use of the same Internet protocol that enables their communication and information exchange (Sula et al., 2013). There is no universal definition of the Internet of Things. All of the definitions have a common feature based on the fact that the first version of the Internet concerns data created by people while the subsequent version concerns data created by things. One of the possible definitions of the Internet of Things is as follows: "An open and comprehensive network of intelligent objects that are able to automatically organize, inform, and facilitate data resources, reactions, and activities in the face of situations and changes in the environment" (Somayya et al., 2015, p. 165). The Internet of Things can be treated as a global network that enables communication between people and objects (Aggarwal & Das, 2012; Sąsiadek & Basl, 2018, p. 189).

The discussed definition category occurred for the first time in the United States of America in 1999, and its first use is said to have been done by Kevin Ashton, an expert in digital innovations. The term was used during the (www.aerogearboxinternational.com/) marketing presentation in which it was underlined that computers (ergo, the Internet) depend near-totally on people and information added to the network by people. According to Gartner, 8.4 billion "things" were connected with the Internet in 2017 (excluding laptops, computers, tablets, and mobile phones). The number is expected to increase to 20.4 billion Internet of Things items by 2020 (Gartner, 2017; Kassab et al., 2019, p. 2).

The Internet of Things focuses mainly on the automatization of most of its processes (in private, public, or professional lives). The purpose of the Internet of Things is to improve different spheres of our life and better manage space and data (whose amounts are increasing at a blistering pace). This concerns housewares, for example, as well as all devices found in a smart home (consumer electronics, central heating, lighting, counters, and intrusion-detection systems).

The definition of the Internet of Things can be given in three aspects:

1. Technological—"IoT means wired and cordless networks of devices marking with independent (not demanding human's engagement) activity in the range of obtaining, facilitating, and processing data or coming in interactions with the environment under the data thumb" (IoT in Polish economy . . . 2019). This means that IT systems and telecommunication networks marked with a high level of dispersion are used in every economic, social, or scientific field of activity to the creation of intelligent control and measuring, analytical systems, or controls.
2. Architectural—"IoT is a concept of IT architecture that enables the cooperation (interoperability) of various IT systems supporting various field applications" (IoT in Polish economy . . . 2019), including the following elements: equipment, communication, software, and integration.
3. Business—"IoT is an ecosystem of business services using objects ready for information gathering and processing (interactions) connected to a network and providing interoperability and synergy of applications" (IoT in Polish economy . . . 2019), whereby the network enables identifying significant situations and events and quickly reacting to immediate optimization or precise personalization orientation concerning clients, the environment, products, and processes.

The Internet of Things concept is not fully recognized in the aspect of business potential. Its development is very intensive nowadays because of the high speed of IT development. Heretofore, independent automated production companies transform into completely automated and optimal productive environments. So-called intelligent factories are established that can exchange information independently with the use of Internet communication protocols, react to potential mistakes in the present time, and adjust to the changing needs and expectations of consumers. Intelligent factories provide competitive products simultaneously (Sąsiadek & Basl, 2018, pp. 189–190). Most experts claim that Industry 4.0 and the Internet of Things offer us a great chance for improving development and competition.

Nevertheless, it is worth mentioning that the concept of the Internet of Things demands many modern solutions and information exchange such as the following:

1. It is necessary to have facilities equipped with various sensors/detectors such as motion detectors, temperature sensors, GPS systems, and vibration pickups. The following devices are necessary to receive a signal, transform it, and cause a specified reaction (e.g., intelligent traffic lights): smartphone, tablet, computer, or other devices that does the activity automatically.
2. Channel of communication is necessary—data transmission (technologies enabling the transmission of information between two elements) beginning from the simplest to the most popular ones (Wi-Fi, Bluetooth, NFC, or Z-WAVE).

People should be aware that IoT development has many limitations, such as the following:

1. Technical barriers—electric supply; in spite of the fact that the devices are being equipped with stronger and stronger batteries, sooner or later, they will need to be plugged in or changed.
2. Architecture of used data transmission systems—it becomes necessary to use new decentralized data transmission systems that additionally guarantee a high level of security (e.g., Blockchain).
3. Lack of uniform standards concerning information security and privacy.
4. Social barriers (especially concerning seniors)—reluctance to utilize new technologies, inability to use new technologies; unclear benefits from the use of new technologies.
5. Legal barriers—confidentiality and privacy are crucial aspects for both private people and organizations (companies) using IoT solutions. Technological solutions are not enough for all user security. Regulations and rules of law must be adjusted to the new reality

before IoT becomes standard fare, not only on the domestic market but also abroad.

6. Consciousness is also a huge barrier to IoT adaptation—to be precise, unconsciousness about how many advantages can be brought due to using the Internet of Things is the biggest problem nowadays.

It has to be emphasized that all of the barriers and limitations of the Internet of Things are in the minority when compared to the advantages. Among the most important benefits of IoT systems are M2M communication, automation and control, information, monitoring, time and money savings, etc. All of the advantages result in a better quality of our lives because all of the applications culminate in increased comfort, convenience, and better management. Among the disadvantages of the Internet of Things, one can enumerate the following:

- Lack of international standard of compatibility;
- Risk of losing privacy increases rapidly;
- Information security is prone to attack by hackers;
- Lesser employment of menial staff;
- Technology takes control of life, etc.

We live in times when technology is spread all over the world, and all of the disciplines and spheres of our lives are engulfed by it. We cannot avoid it, but we can decide to what extent we allow the technology to intrude into our lives. It is worth underlining that the IoT concept is more often implemented by organizations and companies than by private people. This differs from country to country because of technological advancement, infrastructural barriers, ways of thinking, etc. Therefore, the state of the preparations of IoT implementation is very differentiated in various countries and branches or even in individual companies.

Poland belongs to both the most industrialized countries in the European Union and the biggest countries with a high speed of added value in the productive sector. The openness of Polish companies to digitization transformations influences a high level of IoT adaptation on the Polish market. There are still too few companies offering IoT solutions in Poland as compared to the demand for these kinds of solutions. Hence, the creation of suitable conditions for these kinds of companies in Poland has a crucial meaning for the whole Polish economy.

Companies That Fall Into Industry 4.0 Concept in Chosen European Countries

The Internet of Things market develops very fast what can be observed in many companies and institutions.

Aero Gearbox International conducts its activity in compliance with Industry 4.0. The company is marked with an integration of intelligent machines, systems, and implementation of changes in its productive processes. All of these things lead to productivity enhancement and the possibilities of elastic changes in stocks. Aero Gearbox International is a joint venture partnership created by two leading international high-technology aerospace companies: Rolls-Royce plc and Safran Transmission Systems. The company works on designing, developing, producing, and supporting the accessory drive train (ADT) transmission system for all future Rolls-Royce civil aerospace gas turbine engines. The company's vision is to create "optimized accessory drive train solutions for the future." The activity of Aero Gearbox International concerns three key areas:

1. Design and development—using proven technology, they are integrated within both their customer projects and supply chain. They are uniquely positioned to deliver ADT designs optimized to the engine architecture.
2. Manufacturing and assembly—competitive production company working on ADT based on the highest indicators of cost-effectiveness, quality, and deliveries.
3. Services and MRO (technical service, repairs, and renovations)—experience based on repairs and renovations during flight as well as on engineering knowledge toward the development of innovative new techniques to improve service performance and reliability.

Aero Gearbox International is located in four countries:

- France—main seat in Colombes;
- Great Britain—with an agency in Derby;
- Germany—with an agency in Dahlewitz;
- Poland—with an agency in Ropczyce.

Aero Gearbox International was established in 2015. The company constitutes one team that does not depend on the agency's location. They pursue the use of the best practices implemented by the parent agencies. They operate on their own internal CRADLLE system, which provides services across the IT security domain; this helps them achieve a coherent and comprehensive strategy for protecting their corporate assets. The company is equipped with computerized numerical control—equipped with a microcomputer that can be interactively programed at will. The CNC enables the fast, precise, and repetitive production of complicated shapes. Each product must be created separately by the CNC system. Aero Gearbox International is competitive on the market thanks to new equipment that is provided by reputable producers. Hence, the company provides their employees with the newest equipment in this branch that

Table 8.1 Fields of productive machinery maintenance

Fields of productive machinery maintenance	*Range*
Independent maintenance of machinery's motion	• Everyday cleaning, preservation, complementing • Suggestions for improvement • Reports concerning changes
Proactive preservation	• Periodic preservation and monitoring of usage • Calibration and adjustment • Exchange of grease products
Preventative preservation	• Periodic measurement of the degradation level • Monitoring of the technical state of components • Preventative exchange of components

Source: own elaboration on basis of an in-depth interview with Aero Gearbox International

is available on the market. Within the total productive maintenance program, it is possible to monitor all of the activities and provide maximum effectiveness and reliability.

The homogeneous cyber-physical system used in the company is a combination of equipment elements, software, and communication. The machines become more and more intelligent, communicate with each other, and use common data. Moreover, the company is focused on the personalization of services and tailoring. Aero Gearbox International makes the same products for different clients, but there are various standards and requirements. The main client (Rolls-Royce) uses an integrated result card to assess the company as a supplier. The effectiveness is measured in percentages. The following issues are included in the new ways of work and roles in the company:

- Modern machines;
- Implementation processes;
- Control of materials;
- Own laboratory, toughening;
- Pressure washers.

There are four Aero Gearbox International agencies in the world (as mentioned earlier). Nevertheless, all of the agencies fall into Industry 4.0 and its assets as follows:

- Possibility of personal development;
- Advanced machines improving company's prestige;

- Lower production costs;
- Faster delivery;
- Optimization of production;
- Easy communications with clients thanks to fast delivery;
- Improvement of company image;
- Better quality of products;
- Quality control, preventative activities.

Industry 4.0 enables more-terminable deliveries due to which one can observe increases in production elasticity and effectiveness as well as faster reactions to changing client needs and expectations. Aero Gearbox International builds and develops its activity to reach the highest position in the aviation market. Client satisfaction, personal and professional development, new technologies, best practices, and long-term objectives—these are elements of the Aero Gearbox International system. The key objective is realized declined through four major strategic themes:

- Strategy and environment;
- Effective processes;
- Qualified employees;
- Improvement and changes.

Additionally, Aero Gearbox International implements big data; this is also a key activity in the company that influences Industry 4.0 due to the comprehensive assessment of data as well as the management systems realized in various companies.

Another example of the entrepreneurship that falls into the Industry 4.0 concept (with the use of IoT) may be the Polish family company Greinplast LLC (www.greinplast.pl/), which has developed dynamically on the structural chemistry market since 1997. The production is fully automated, which minimizes the risk of downtime due to human resource reasons. The production lines are managed and programed by specialists who have knowledge of the products' recipes; these recipes are determined in the laboratory. The production process starts when specific values are in the system. The line uses its "intelligence" through weight systems, dosing settings, and control cabinets. It additionally provides lots of other information such as the amount and weight of the produced articles, quality of the ingredients, production changes, and diagnostic errors. All of the information above-mentioned enables the employees to respond to possible irregularities or failure removal in a fast way. The production lines "communicate" with the personnel in this way; with the indicated error code, it is possible to minimize diagnostic time. The production process can be planned in advance and stimulated due to demand. It decreases the risk of producing too many articles, for example, and it makes it possible to better manage stores or a store's cost cutting (which ultimately results in more benefits). Additionally, the line uses the collected data and optimizes the production process. The line consists of several machines that must communicate with each other to provide the continuity and efficiency of production. This is defined as a cyber and physical process connection with highly advanced technology. The effect of the system is the unification of the real world of machines with the virtual world of the Internet (the Industrial Internet of Things). Few employees have remote access to machines, which enables them to download new recipes and improve the old ones with the purpose of building an institution. It is also possible to correct prospective data errors and diagnose defects at a fast pace. Each of the employees has access to the production line from all over the world. The service also has additional access to the machines. In case of a problem, the service can respond very quickly and redress the damage in a remote way. This results in reducing the stoppage of temporary productions and minimizing the risk of a production plan's non-execution. It is worth underlining that the production line settings in the Greinplast company are at a very high level because one factory is enough for realizing product demand in Poland as well as abroad.

It is also worth analyzing companies that implement Industry 4.0 in non-European Union countries. The analysis is based on a company's functioning in Ukraine where employees wrote software to manage display advertisements in 2012. They unfortunately had troubles with deliveries and sales of their products. Then, Roman Krawczenko (a representative of the company and founder of the IoT Hub accelerator laboratory) made a decision about going to the United States in order to gain new experience in software creation. After his return, he created a new line of displays along with other experts. This resulted in the fact that he became not only a programmer but also an entrepreneur within a few years. He realized that there is a necessity for hybrid model creation in Ukraine: a kind of partnership in which they invest in a project and aim the team toward common goals. This is key to new production. There are many such start-ups in the United States. These opportunities are much rarer in Ukraine, so it is very important to educate specialists in this area. This way of education and activity was chosen by the mentioned team of the company; it resulted in an interesting formula of getting external funds such as crowdsourcing, crowdfunding, and co-investing. It was on the Kickstarter crowd-funding platform (which is not cheap but is based on image, loyalty, and global society). The solutions were implemented out of Ukraine, and the company uses technology as the Internet of Everything. The Pix project (unnamed backpack with a colored display that enables expressions in the form of individual icons that can be designed individually) is an example of the solution. In addition, the company is engaged in the Senstone project—a dictating machine that transforms voice to text. It is worth mentioning that the dictating machine and backpack markets already exist, but the company merely combines these objects (https://aiconference.com.ua).

Effective projects demand people with similar expectations and interests that are going to support each other. This kind of cooperation was developed in Ukraine.

The use of the Internet of Things associates mainly with production and industry. We rarely think about IoT in the context of daily life. A good example of entrepreneurship that acts due to the Industry 4.0 concept offers technologies that improve quality of daily life may be the Bosh Group, whose units function in around 60 countries all over the world. The Bosh group falls into the concept of the virtual and real worlds connecting, creating all of the pictures from an integrated world and working on technologies to improve the quality of everyday life.

Among the products that introduce Bosh's clients to the IoT world, one can enumerate them as follows (www.bosch.de/unser-unternehmen/bosch-gruppe-weltweit):

1. Mowing robot Indego with intelligent navigation system "LogiCut" that not only cuts the grass but can also determine the ideal cutting program (it will calculate the optimal cutting time on the basis of information analysis concerning grass growth, the current weather forecast, and data concerning the surface of the lawn).
2. Intelligent central heating management system—Bosh EasyControl, whose basis is a CT200 regulator and the Internet due to which the system communicates wirelessly with mobile devices.
3. Home Connect platform—as a houseware navigation center that enables the devices to communicate with the Internet (through the Wi-Fi network) as well as with other devices and smartphone apps. The apps' ecosystem is being created in such a way as to enable people to navigate greater numbers of devices at home. The Home Connect system is used in advanced-technology clothes dryers (and many other devices), whereby having a dryer and washing machine that are equipped with Home Connect technology at their disposal means that there is no need to choose a drying program. The dryer can then choose the optimal drying program (on the basis of analyzing the weight, ending the program, and characteristics of the spin cycle).

It is worth mentioning that Bosh started to digitize its production at the end of the 20th century, while it started to implement dreams concerning an intelligent and (almost) independent factory in 2015 (turning on the first assembly line to the semi-automatic production of hydraulic distributors). The model Bosh Rexroth factory in Hamburg works on a production line composed of integrated modules in Bosh Production Systems (BPS). The use of radio-frequency identification technology (RFID) enables workstations to recognize a product's type in order to choose suitable materials and production processes automatically. The employees of the factory control the correctness of the technology by working with the ActiveCockpit system that presents the data of machines and processes online. Therefore, it is possible to correct prospective irregularities swiftly.

There are more and more entrepreneurships that fall into the Industry 4.0 concept. Many companies evolve in this direction, even subconsciously, because they participate in complicated logistical systems consisting of production partners, component providers, suppliers, and sales on the open European market. One can enumerate many benefits from the implementation of intelligent production (among others: increase of productivity through the possibility of the better use of assets and minimization of temporary stoppages; the better planning and monitoring of production processes; the possibility of personalized products due to client preferences with a concurrent minimization of a production's marginal costs; optimization of production costs due to wastage identification and cost monitoring; better adjusting to market demands, faster responding to changes).

Conclusions

The Industry 4.0 concept results in many changes not only in production processes but also in structures and companies' various fields of activities (for example, in management, logistics, communication, etc.). In the range of technology, the concept concerns many technologies such as the Internet of Things, cloud computing, big data analysis, and artificial intelligence as well as augmented reality and cooperating robots.

On the other hand, the Internet of Things (according to which, devices can gather, process, and exchange data thanks to communication networks—especially the Internet) is not fully recognized and analyzed in the aspect of business potential. It has developed quite intensively in recent years, which is connected with the fast development of network technology.

To sum up all the considerations and analyses made in this chapter as well as those accessible in the literature of the subject, one should conclude that the Internet of Things constitutes a great opportunity for the development and improvement of competitiveness. Nevertheless, it has to be emphasized that the state of its implementation is differentiated in various countries, branches, and even individual companies. Eastern Europe ranks lower than Western European countries and North America according to the number of companies offering IoT solutions. It is also very significant that most of the Eastern European companies offering IoT solutions are located in Poland, which has a great IoT potential occurring from a high level of adaptation and high demand. Despite the high level of IoT potential in Poland, its direction and development pace depends on the adaptation processes of the producers and service providers.

Improving competitiveness and increasing added value for consumers is a big challenge both for European countries and non-European countries in the context of IoT, which touches our daily lives and everything that is connected. Security should be a priority in the strategies of producers whose products go to clients. Consumers are in need of solutions built with security in mind and that protect their lives and private information. There is a need for better regulations due to a lack of product certifications to help consumers in differentiating brands and products (or services) with respect to their security states.

These recommendations refer to companies that fall into the IoT idea and should be focused on their clients' security (especially IT security) because security management in IoT environments is of fundamental importance (Ren, 2011; Chen & Helal, 2011; Roman et al., 2011; Zhou & Chao, 2011). According to some researchers (Apthorpe et al., 2018), the proliferation of Internet of Things devices for consumer "smart" facilities raises concerns about their privacy. Moreover, providing security is essential for IoT development, which concerns the creation of suitable procedures of standardization and certification being an effective and economic confirmation of IoT technology quality. It is also a mechanism that shows suitable standards in the aspect of cybersecurity and communication. The solutions can also be recommended to global corporations investing in the development of their own products and services within the Internet of Things concept. The authors also believe that the recommendations should also be taken into account by units and organizations working on consumer security against discrimination in the areas of service and the responsibility of product providers as well as the possibilities of using data collected in public domains.

References

Aggarwal, R., & Lal Das, M. (2012). *RFID security in the context of "internet of things"*. First International Conference on Security of Internet of Things, Kerala, August 17–19, pp. 51–56.

Apthorpe, N., Shvartzshnaider, Y., Mathur, A., Reisman, D., & Feamster, N. (2018). Discovering smart home internet of things privacy norms using contextual integrity. *Proceeding of the ACM on Interactive, Mobile, Wearable and Ubiquitous Technologies*, 2(2).

Ashton, K. (2010). That 'internet of things' thing. In the real world, things matter more than ideas. *RFID Journal*. www.rfid.journal.com/articles/pdf?4986

Bosch weltweit. *Die Bosch-Gruppe im Überblick*. www.bosch.de/unser-unternehmen/bosch-gruppe-weltweit/

Challenges and solutions for the digital transformation and use of exponential technologies. https://www2.deloitte.com/content/dam/Deloitte/ch/Documents/manufacturing/ch-en-manufacturing-industry-4-0-24102014.pdf

Chen, C., & Helal, S. (2011). *A device-centric approach to a safer internet of things*. International Workshop on Networking and Object memories for the Internet of Things, Beijing, China, pp. 1–6.

Gartner. (2017). *The Internet of Things (IoT) is a key enabling technology for digital businesses*. www.gartner.com/technology/research/internet-of-things/

Greinplast. www.greinplast.pl/

Guo, Y., Liu, H., & Chai, Y. (2014). The embedding convergence of smart cities and tourism internet of things in China: An advance perspective. *Advances in Hospitality and Tourism Research (AHTR)*, *2*(1), 54–69.

Industrie 4.0 & Chancen und Herausforderungen der vierten industriellen Revolution, p. 5. www.strategyand.pwc.com/de/de/studie/industrie-4-0.pdf.

Internet of Things in Ukraine: History of Iot Market by Roman Kravchenko, Iot Hub Founder. https://aiconference.com.ua/en/news/internet-veshchey-v-ukraine-istoriya-sozdaniya-iot-rinka-ot-romana-kravchenko-osnovatelya-iot-hub-95075

IoT w polskiej gospodarce. (2019). *Raport grupy roboczej do spraw Internetu Rzeczy przy Ministerstwie Cyfryzacji* (p. 5). Ministerstwo Cyfryzacji.

Kassab, M., DeFranco, J., & Laplante, P. (2019). A systematic literature review on internet of things in education: Benefits and challenges. *Journal of Computer Assisted Learning*. www.researchgate.net/publication/333643709_A_Systematic_Literature_Review_on_Internet_of_Things_in_Education_Benefits_and_Challenges

Lasi, H. P., Fettke, P., Kemper, H. G., Feld, T., & Hoffmann, M. (2014). Industry 4.0. *Business & Information Systems Engineering*, *6*(4), 239.

Maple, C. (2017). Security and privacy in the internet of things. *Journal of Cyber Policy*, *2*(2), 155–184.

Ren, W. (2011). QoS-aware and compromise-resilient key management scheme for heterogeneous wireless internet of things. *International Journal of Network Management*, *21*(4), 284–299.

Roman, R., Najera, P., & Lopez, J. (2011). Securing the internet of things. *Computer*, *44*(9), 51–58.

Sarma, S., Brock, D., & Ashton, K. (2000). *The networked physical world: Proposals for engineering the next generation of computing, commerce, and automatic-identification* (pp. 1–16). White Paper Auto-ID Center (MIT-AUTOID-WH-001). John Wiley & Sons, Inc.

Sąsiadek, M., & Basl, J. (2018). Świadomość i poziom wdrożenia koncepcji Przemysł 4.0 w wybranych polskich i czeskich przedsiębiorstwach. In R. Knosala (Ed.), *Innowacje w zarządzaniu i inżynierii produkcji* (pp. 189–190). T. 2, Oficyna Wydaw. Polskiego Towarzystwa Zarządzania Produkcją.

Somayya, M., Ramaswamy, R., & Tripathi, S. (2015). Internet of Things (IoT): A literature review. *Journal of Computer and Communications*, *3*, 164–173.

Sula, A., Spaho, E., Matsuo, K., Barolli, L., Miho, R., & Xhafa, F. (2013). *An IoT-based system for supporting children with autismspectrum disorder*. 2013 Eighth International Conference on Broadband and Wireless Computing, Communication and Applications, Washington, DC, pp. 282–289.

Weber, R. H. (2010). Internet of things: New security and privacy challenges. *Computer Law and Security Review*, *26*(1), 23–30.

Xu, L. D., & Duan, L. (2018). Big data for cyber physical systems in industry 4.0: A survey. *Enterprise Information Systems*, *13*(2), 148–169.

Xu, L. D., Xu, E., & Li, L. (2018). Industry 4.0: State of the Art and Future Trends. *International Journal of Production Research, Francis & Taylor*, *56*(8), 2941–2962.

Zhou, L., & Chao, H. C. (2011). Multimedia traffic security architecture for the internet of things. *IEEE Network*, *25*(3), 35–40.

9 Cyber-Physical Logistics Platform—Toward Digitalization of Printing Company

Jerzy Duda, Robert Goncerz, Iwona Skalna, Daniel Kubek, Katarzyna Rybicka, Paweł Więcek, Tomasz Derlecki and Radosław Puka

Introduction

The printing industry has a long tradition in Europe; it is an important employer that provides jobs for highly skilled and qualified staff. The sector is currently facing challenges related to changing habits, shifts in alternative ways of communications, and an increase in competition in the printing market. Online shopping and the paperless revolution have reduced the number of orders from traditional customers and, hence, caused the need to look for new ones. In addition, customer requirements are constantly increasing; for example, they want to be able to make flexible decisions on the print run and volume with a comprehensive tailored-made supply chain to reflect real-time advertisement needs. There is also a growing environmental awareness within the industry to reduce the negative impact of their operations on the environment through waste reduction, recycling, and the reduced use of energy, water, and chemicals, strengthened by European Union law.

To meet all these challenges, printing companies must undergo a digital transformation, tailored to their needs and possibilities, that should embrace all major business processes. This transformation can be achieved with help from cyber-physical systems, the Internet of Things, big data, cloud computing, and industrial wireless networks, which are the core of the Fourth Industrial Revolution (Industry 4.0). In this chapter, we present the most important components of the cyber-physical logistics platform (CPLP) that we have developed for a printing company to improve its sustainability and optimize the key processes in the company's logistics chain. The platform uses the CPS approach to integrate the supply, production, warehousing, and distribution processes of a company with computational and network processes. The following sections describe the general architecture of the platform as well as its layers.

DOI: 10.4324/9781003186373-9

Therein are also described the challenges related to the implementation of the platform in the area of internal logistics, warehouse management, production, and distribution to the final customer. We do believe that the presented solution will allow to increase the company's efficiency and strengthen its strategic advantage in the future.

Cyber-Physical Systems for Industry 4.0

A cyber-physical system (CPS) is a core component of Industry 4.0 (Xu et al., 2018). CPS integrates dynamic physical processes with computational and network processes. The network of industrial and embedded computers, controllers, and wired and wireless sensors monitor and control the physical processes with feedback loops, where the physical processes (the so-called physical components of CPS) affect the computations and cyber models (the so-called cyber components of CPS) and vice versa. The cyber models create virtual copies of the physical processes to allow interaction with them. These virtual copies (cyber-representations) are created based on the digital data and information about the physical components. In this sense, CPS can be seen as a collection of transformative technologies for managing interconnected computational and physical capabilities (Trappey et al., 2016). However, it should be emphasized that CPS is not the union of physical and cyber components but rather the intersection of them. Therefore, it is necessary to know the nature of this intersection and understand how the components interact (Lee & Seshia, 2014).

Examples of CPS include smart grids, autonomous automobile systems, medical monitoring systems, industrial control systems, and robotics systems. Various manufacturing industries implement CPS to advance production, distribution, transportation, service, and maintenance in the manufacturing process (Tao et al., 2018). The implementation of CPS in the field of production management has been formalized over the past few years (Monostori, 2014) under the term of cyber-physical production systems (CPPSs). According to Rudtsch et al. (2014), the main benefits that can be expected from the generalization of CPPSs are as follows: (i) the optimization of production processes, (ii) optimized product customization, (iii) resource-efficient production, and (iv) human-centered production processes. Cyber-physical logistics systems (CPLSs) implemented in the context of manufacturing are the results of CPPSs applied to logistics (Pujo & Ounnar, 2018).

Cyber-physical systems are closely connected with the concept of a Digital Twin (DT). Like CPSs, DTs integrate the cyber world with the real world through real-time communication; however, virtual reality and the model of the behavior of physical entities are their fundamental tools (Boschert & Rosen, 2016). DTs enable the prediction and optimization of mapped objects by real-time simulation and bi-directional communication based on real-time data from various sensors and controllers. As compared to CPSs, DTs provide a realistic mirror image of real objects by simulating their behaviors and functions. The architecture of CPS does not provide exact mirror models of physical objects but strongly emphasizes the computing and communication technologies in cyberspace. Therefore, CPS may affect more than one entity in a complex system. (Tao et al., 2019).

However, there are no strict rules for the construction of CPSs. Lee et al. (2015) proposed a five-level architecture (5C) for developing and deploying a CPS for manufacturing applications. The proposed architecture is conceived in principle as guidelines for the construction of such systems; that is, how to transform physical components into digital components based on data and information, how to process this data and information so that it is possible to make optimized decisions, and finally to achieve the self-regulation of the whole system (both its physical and cyber parts). The smart connection level (Level 1) represents the physical components. The data for this level can be gathered from sensors as well as from SCADA (Supervisory Control and Data Acquisition), MES (Manufacturing Execution System), or even ERP (Enterprise Resource Planning) systems. Lee et al. (2015) emphasized the role of the appropriate speed of data transfer from these sources and the issue of the proper selection of sensors and interfaces (see Section 4.1). Levels 2–4 represent the cyber components of CPPS (cf. Monostori, 2014). The data-conversion level (Level 2) primarily focuses on extracting useful information from the data that can be used to determine the health condition and performance of machines or to predict their degradation levels (for example). In a narrow sense, the cyber level (Level 3) consists of virtual digital copies of machines and other physical objects (cyber-machine model, digital twin); in a broader sense, however, it also includes analytical instruments that can find patterns in the data as well as similarities to historical data. The cognition level (Level 4) generates in-depth knowledge about the monitored system; it is responsible for supporting the decision-making processes (which include, inter alia, integrated simulation, and synthesis followed by supportive visualization of information for humans). Finally, the configuration level (Level 5) realizes the feedback from the cyberspace to the physical space—it applies the corrective and preventive decisions taken in the cognition level to the monitored real-time system. Various aspects related to the architecture of CPSs are discussed in detail by Liu and Zhang (2015).

Similar layers, though in the broader context of comprehensive Industry 4.0 systems, can be found in a three-dimensional architecture described as the Reference Architectural Model Industrie 4.0 (RAMI 4.0). The architectural axis of this model lists six layers: (1) Assets—physical things in the real world, (2) Integration—transition from real to digital world, (3) Communication—access to Information, (4) Information—necessary data, (5) Functional layer—functions of the assets, and (6) Business layer—representing organization and business processes (Pascual et al., 2019).

CPSs can also be considered to be the technological foundation of logistics and supply chain management (Klötzer & Pflaum, 2015), which are a core pillar in the value chain for manufacturers and retailers. Several concepts of using CPSs in logistics can be found in the literature. The simplest way is to use an existing CPS and enrich it with basic big data analysis models for monitoring and managing warehouses, flow, allocation, and distribution. One such model (a model for training purposes) was presented in Seitz & Nyhuis (2015). A similar idea was presented by Frazzon (2005); however, systems are proposed that cover much more complex logistics processes in real applications. Panetto et al. (2019) proposed a digital cyber-physical supply chain framework that consists of three parts: (i) a physical supply chain, (ii) supply chain and operational analytics, and (iii) a cyber-supply chain. The physical supply chain provides data for the cyber supply chain that models all the main processes. The data for the procurement process can be obtained from ERP and APS systems, including e-procurement and supply visibility control systems. The manufacturing process can provide data from RFID, GPS, different sensors, robotics, and virtual or augmented reality. The logistics processes provide data from Tracking and Tracing (T&T) systems as well as transactional systems. Finally, the sales process generates lots of transactional data at the points of sales. The data collected for the cyber-physical chain is then transferred to a supply chain and operations analytics module that is responsible for the following:

- Descriptive and diagnostic analyses;
- Predictive simulation and prescriptive optimization;
- Real-time control;
- Adaptive learning.

Most recently, Kong et al. (2020) presented a cyber-physical e-commerce logistics system built on cloud service architecture. The first level of the system is provided as infrastructure-as-a-service (IaaS) and contains three layers: (i) a physical layer (human resources and logistic facilities), (ii) a perceptual layer that includes sensing technologies and industrial wearable technologies; and (iii) a control layer in the form of a mobile gateway operating system that works as a gateway for the higher layers. The second level works as platform-as-a-service (PaaS) and has two key components: (i) an intelligent coordination system (iCoordinator) as the core technology in the execution layer, which is responsible for the execution of the synchronized order fulfillment process and (ii) an intelligent synchronization system (iSync) that works in the scheduling layer to solve any synchronization problems. The third (top) level is the analytics layer that works as Software-as-a-Service (SaaS) and provides three services: (i) multidimensional visualization; (ii) virtual space management; and (iii) value-added data analytics that stores models for process optimization from the supply side to the demand side.

However, the current literature does not present any advanced complex cyber-physical logistics system that can be used in production practice to control and manage all important logistic processes. The following sections present an attempt to fill this gap by proposing a cyber-physical logistics platform that aims to optimize and partially automate logistics processes in a web and sheet-fed offset printing enterprise.

Architecture of Cyber-Physical Logistics Platform

The developed cyber-physical logistics platform (CPLP) must achieve two main business goals: reducing logistics costs and improving the production process (including its efficiency and reliability). These goals can be further specified for the key processes as follows:

- Improve the efficiency of key logistics operations at the factory;
- Improve coordination between changes in production buffers and transport (reduce buffers);
- Reduce energy consumption;
- Reduce the number of logistics operations, which has an indirect impact on damage to materials or semi-finished products (for example);
- Reduce material grace time due to the better monitoring of the warehouse climate;
- Obtain information (as complete as possible) about the state of the key production processes and support processes—better and faster response to emergencies;
- Eliminate downtime resulting from delays in the delivery of materials or semi-finished products;
- Plan production based on real and predicted machine stock and inventory levels.

For simplicity, the architecture of the CPLP (see Figure 9.1) will be presented with only three basic layers: physical, communication, and digital. The communication layer, however, corresponds in the RAMI 4.0 model to both the data and communication layer (everything, including data from ERP, WMS, and TMS systems is connected by a dedicated data bus), while the digital layer corresponds to the information and functional layer (cognitive and configuration layer in the 5C model).

The main assumption for the developed CPLP architecture was the use of available data sources regarding the physical objects of the production system, its supplementation, and (where necessary) the expansion of the

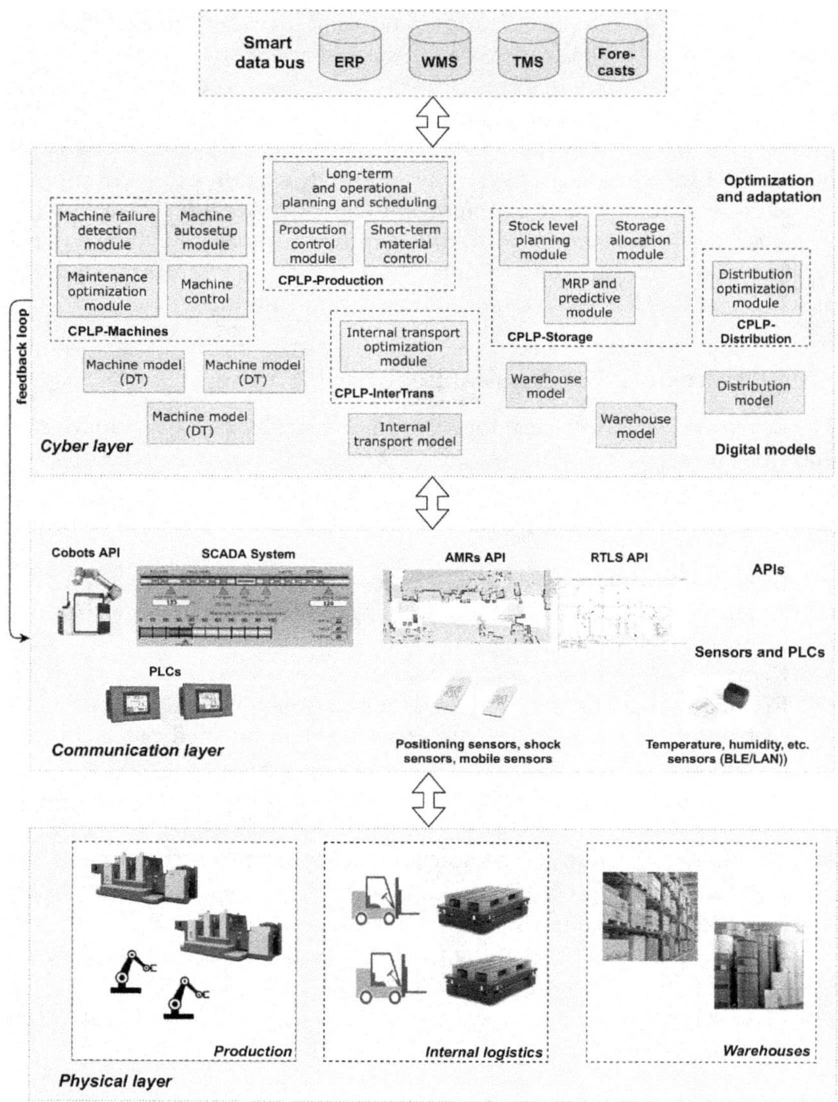

Figure 9.1 Architecture of cyber-physical logistics platform (CPLP).

Source: authors' own work

object-sensing system (machines, production halls, warehouses, internal transport). An important assumption was also the gradual introduction of new means of production and transport equipment with autonomous systems controlling their work (e.g., autonomous forklifts, cobots). Therefore, the physical layer of the CPLP consists of physical objects (including traditional means of production, such as printing machines, binding machines, and others requiring human operators), autonomous objects, as well as shop floors and warehouses (understood as a source of data from the existing systems and sensors). The variety of the physical objects is also reflected in the communication layer. The main medium of data transmission is Ethernet, as it provides adequate speed and (above all) stability under production conditions. Ethernet networks are used for SCADA systems and RTLS as well. Wireless connections—based on Wi-Fi and Bluetooth Low Energy (BLE)—are used when possible (e.g., no interference occurs, no high bandwidth is required—primarily for different types of sensors) or when required (e.g., for communication with autonomous robots).

The cyber layer includes digital models of the machines—each machine has its own model, which primarily uses information from programmable logic controllers (PLCs) and the data from the sensors installed at the machines. In both cases, data transmission takes place via the SCADA system, if possible. The models for the other processes are not real digital twins of the physical objects, but they concentrate on the most important attributes of these objects (necessary for supplying information for the modules of the CPLP): the internal transport model maps primarily current location and the condition of the forklifts (and data from collision sensors, for example). Warehouse model maps primarily the location of materials and semi-finished products (and data from BMS sensors like temperature and humidity). Finally, the distribution model relies on data regarding the status of shipments as distribution is carried out to a large extent by external entities, and therefore, tracking by GPS modules is not used.

At the top of the cyber layer are optimization models that allow for the automatic or semi-automatic control of key logistics processes (e.g., with HMI interfaces). They also allow for the proper control of machines and devices in the physical layer as well as for adjusting the physical objects parameters. In addition, some models in this layer allow for processing the data from the communication layer in combination with the data from enterprise information systems (ERP, TMS, WMS, forecasting).

Physical Layer

The physical layer includes physical objects that are modeled within the CPLP. These physical objects can be divided into the machinery that is directly used for production (e.g., printing machines, cutting machines, bookbinding machines) and the auxiliary machines that are used to automate selected manual operations, such as packaging or foiling (e.g., cobots). The next group is the devices that are used for internal logistics (mainly forklifts). Traditionally, companies have used forklifts operated by a person who receives transport commands via a headset. However, to support the implementation of the CPLP, a company must purchase autonomous vehicles (the AMR type). Such vehicles are operator free and follow routes that are provided digitally (via API or a dedicated application). It is also always possible to read the current position and status of a forklift. To obtain the complete digital image of their forklifts, a company must purchase BLE sensors that are used to construct a real-time locating system (RTLS), which allows them to read the position of all their forklifts with some accuracy.

Finally, the third major group of physical objects being a part of the lowest CPLP layer are warehouses for raw materials (including main raw material—paper), semi-finished products, and finished products. These warehouses have been equipped with additional sensors that are a part of the building management system (BMS) and allow for the monitoring of environmental parameters (such as temperature or humidity). The data obtained from the sensors will be used in the CPLP to determine the period during which the paper must remain in the warehouse to achieve the appropriate properties. Additional sensors should also be planned to increase the precision of items location in the warehouses without shelves (e.g., a paper warehouse).

Since our intention is not to delve into the details about the technological processes of the offset printing house, we will mainly focus on autonomous transport in the next section, and we will present an idea of how to manually embed the operated forklifts into the CPLP platform.

Due to the recent development of technology and the prevalent logistics framework, the relevance of digitalized and autonomous transport is growing rapidly. Autonomous transportation inside industrial plants is becoming commonplace, as it is the only way to keep pace with industry needs. Systems of automated guided vehicles (AGVs) often support internal transportation in manufacturing facilities and warehouses. Until recently, AGVs were the only option for automating internal transportation tasks; today, however, AGVs are being challenged by the more sophisticated, flexible, and cost-effective technology of autonomous mobile robots (AMRs). AMRs navigate via maps constructed by their software on-site or via pre-loaded facility drawings. Such robots are equipped with advanced navigation technology to ensure safe and efficient movements. Two laser scanners mounted on the front and back provide 360-degree field of vision for spotting objects up to 8 meters away. Cameras on the front allow the robot to see and recognize objects above the floor (up to two meters), while proximity sensors in each corner increase the robot's ability to see pallets and other obstacles at lower heights ('MiR', 2020). An AMR can be controlled by a dedicated API or using a special management software. The latter allows for the registration of all robots followed by the automatic selection of a robot for a given task based on its location and availability. Dedicated lifts are used for automated pallet pickup. The operator delivers pallets to the lift and calls a MIR robot using the operator's dashboard on the HMI mounted on the lift. Sensors mounted on the pallet racks indicate their states (free or occupied) via PLCs connected to the industrial network.

Communication Layer

The communication layer combines hardware and software with the purpose of collecting and processing the data coming from the physical layer. The hardware includes equipment, sensors, computers, and other components (like PLCs), while the software concerns working methods, programs, abstract logic, and other basic instructions (Venkata Sundeep, 2013). The real-time monitoring of an entire industrial process needs a reliable network of sensors integrated by fail-free protocols and applications. Therefore, monitoring the architecture (including the metering infrastructure) is of critical concern. In the following sections, we present an overview of the available communication technologies for internal transport. We also provide reasons for choosing specific technologies and other elements of the monitoring system being built.

Sensor Network for Internal Transport Supervision

Proper supply-chain management requires the transportation media and loads to be accurately located and tracked in real time (RTLS localization), among other things. External transport can be monitored by using *global positioning systems* (GPSs), which offer some RTLS capabilities. However, GPS signals cannot penetrate most construction materials, leaving indoor facilities inaccessible. Instead, the *indoor positioning systems* (IPSs) enable several location-based solutions including RTLS.

Digital tracking technologies involve algorithms that compute the positions of vehicles using data coming from fixed reference points (also called location anchors), which detect signals from the markers mounted on moving objects. These algorithms differ in terms of the quality and accuracy of the location detection and the frequency of the location updates. In addition, each type of technology forces the implementation of dedicated methods for result analysis, which can entail additional costs.

Using the STEM-DPR method (Benayoun et al., 1971), we have found that Bluetooth LE from technology will be the most suitable for the purposes of the developed CPL platform, as it provides satisfactory precision in determining the locations of objects while maintaining the minimum risks associated with the amount of investment (the arrangement of permanent access points) and the frequency of technical inspections. In the view of further system developments, an additional factor indicating that the selected solution is optimal is the possibility of implementing more sensors installed on transport vehicles. According to our analyses, these are vibration, temperature, and humidity sensors. Supplementary equipment for Bluetooth LE tags with these sensors (especially the addition of a vibration sensor and accelerator) allows one to gain additional functionality, such as the following:

- Registering the environmental conditions of a plant (for the BMS module) and optimizing the recorded data in the CPL platform. The recorded vibrations of a vehicle indicate whether the vehicle is moving or not. The data gathered during a stop is used to create a temperature/humidity map of the plant (longer stops allow us to determine the temperature more precisely near the stopping region) as well as to record the position of the vehicle only when it is moving (based on the indications of the accelerometer).
- Optimizing vehicle operation—registering overloads occurring during the transportation of materials allows to determine the usage of particular vehicles and, consequently, improves fleet efficiency. Additionally, the most and least efficient processes carried out by the vehicles can be identified; on this basis, the developed algorithms can optimize the routes.
- Event recording—algorithms developed within the CPL platform can generate notifications of adverse events involving warehouse vehicle collisions, particularly with structural elements of the plant hall.

The created temperature/humidity map can be used to optimize the work of other systems such as BMS—including managing the environmental conditions of a hall. The optimal temperature and humidity in a printing enterprise can significantly improve the quality of the print as well as the safety of the production process. Otherwise, the serious drawback can appear, such as the following:

- Paper deformation—if the humidity is too low, paper dynamically releases moisture into the environment, which changes the paper web dimensions and causes undesirable deformations to occur.
- Electrostatic induction—processing paper web with humidity that is too low is difficult, as it gains electrostatic charges more easily; this results in the fact that it cannot move freely through the sections of a printing machine.
- Lack of standardization of the printing process—the variable relative humidity of the air does not allow for standardizing the printing process, which leads to an increase in the amount of wastepaper and production time; downtimes and machine calibration times can also be extended.

The need for the continuous registration of a plant's environmental parameters as well as the basic dynamic parameters of the production machines and the supplied raw material was pointed out while creating the technical assumptions for CPLP. The recorded values of the parameters are necessary to create a digital model of the printing machines and investigate the impact of external conditions on the production process and the behavior of individual sections of the technological line.

BMS Sensors and Registration of Production Parameters

Only the automatic monitoring of machine operation and continuous recording of the parameters of the raw materials and media can ensure that production is more efficient. However, the correct digital mapping of the technological processes is only possible if the parameters that should be read and recorded during the process are properly defined. The main areas that are subject to registration and analysis are as follows:

- Technological line—the specific sections and modules of a production machine. The vibrations of machine elements, linear web speed, or rotational speed of a drive train can be used as input for machine-learning methods to adjust the technological parameters to best carry out the production process. Additionally, it is worth recording all downtimes and failures along with their causes, the maintenance work carried out, and its result. It is also worth classifying the causes (e.g., failure, stop, micro-stoppage) and add a detailed definition of each class. Due to all of this, an appropriate strategy for the operation of production machines can be developed.
- Production floor—environmental conditions (temperature, dust, or humidity) are as important for the production process as the process parameters. Particularly noteworthy is the dustiness, which may result not only from the quality of the paper used but mainly from the industry's scheme. The phenomenon of paper dusting results from higher roughness; this has an adverse effect on the quality of printing and significantly increases the consumption of ink. The application of appropriate sensors may also increase work safety in the terms of fire.
- Raw materials—the correct raw material parameters guarantee the repeatability of the production process and the appropriate quality of the finished products.
- Energy carriers—energy sources having certain parameters; they ensure a continuous supply of energy into the production process. The digital modeling of the production machines and production processes shows that the quality of the media delivered is very important for production efficiency and the quality of the finished products. The development of appropriate algorithms is key to optimizing the amount of media consumed depending on the type of product being produced and the external conditions.

The development of digital models of the production line machinery together with the registration mechanism of the above-presented parameters allows for indicating the possibilities of energy recovery and the reduction of energy consumption. It is also expected that the digitization process will reduce the calibration time (make ready process) by predefining production parameters, what in turn will support the energy consumption decrease. Similarly, the automatic failure reporting model and the preventive review mechanism, should improve the efficiency and level of use of a production line and minimize the consumption of utilities and raw materials.

Supervisory Control and Data Acquisition

Taking into account the CPLP and IoT requirements, it can be clearly stated that HMI (human–machine interface) and supervisory control and data acquisition (SCADA) software are critical elements in the process of plant digitization. The impact of this software can be seen in production management systems and IT systems (these can be MES/WMS/CRM/CBM and other solutions used in the enterprise), especially when it comes to obtaining data from local sensors or recorders for physical servers or cloud computing. The construction of a digital twin (DT) and a cyber-physical system (CPS) should be based on a high-class SCADA/HMI system and the Internet of Things (IoT). This is due to the need for a faster and better information flow, interoperability, security, standardization, and especially user experience (along with its technical intuition). The above digital models require large data sets for further advanced analytics and processing, including the use of artificial intelligence. Properly developed SCADA systems allow for the partial or fully autonomous understanding of the process and decision-making; their results serve as data sources for cyber-physical platforms. When choosing the right SCADA-class system, appropriate technological solutions using a distributed architecture and ensuring proper communication with many components of the production line should be considered. It should be noted that the process of supporting all modern technologies (such as virtualization, cloud, mobile devices, touch panels, or data analysis systems) and the presentation of trends is not without significance for such systems.

Cyber Layer

The CPLP we develop uses the CPS approach for the main phases of the company's logistics system that is, supply, production, warehousing, and distribution. The CPLP at the level of optimization and adaptation layer consists of autonomous and cooperating subsystems that are integrated dynamically and adaptively at all levels of the company's logistics (supply, production, storage, and distribution). The synchronization and cooperation of individual resources (human, technical, or digital) from the mentioned subsystems are achieved by mechanisms of functional-informational integration (e.g., embedded systems, network processing) and by using modern methods such as optimization models and machine learning.

The core CPLP logic reflects the logical division of material flow in the company. The top layer of the cyber-physical logistics platform consists of the following integrated key components:

1. CPS for procurement and storage (CPLP Storage);
2. CPS for internal transport (CPLP InterTrans);
3. CPS for the production process (CPLP Production);
4. CPS for the distribution process (CPLP Distribution);
5. CPS for automation, control, and maintenance prediction for machines (CPLP Machines).

Most of the CPSs listed above are designed to have two major functional components: (1) real-time connectivity of the physical world with cyber-space and feedback information from the cyber to the real world and (2) intelligent data management, analysis, and optimization, which creates the cyber world (Lee et al., 2015).

The first of these (CPLP Storage) consists of the following four modules:

1. Digital warehouse model—it stores (static and dynamic) information about the warehouses, for example, current warehouse occupancy, temperature, and moist space distribution in the paper warehouse or the layout of the warehouses.
2. Predictive model for planning material requirements—current and historical data about the resource requirements are the main source for this module. Its main role is to guarantee the high-quality prediction of material consumption for another part of the CPLP, such as the module for stock level planning or the module for production planning.
3. Optimization model for the allocation of the storage space for materials—the goal of this module is to find the optimal allocation of materials in the storage areas, taking the supply plans, current production, and resource requirements into account. The challenging issue is the paper warehouse that utilizes block-stacking storage. This type of storage makes the objective of the mathematical model to minimize the shuffling of rolls.
4. Stock level planning model—the main role of this module is to adjust the stock level of the materials to the variable production requirements.

The conceptual structure of CPLP Storage is presented in Figure 9.2.

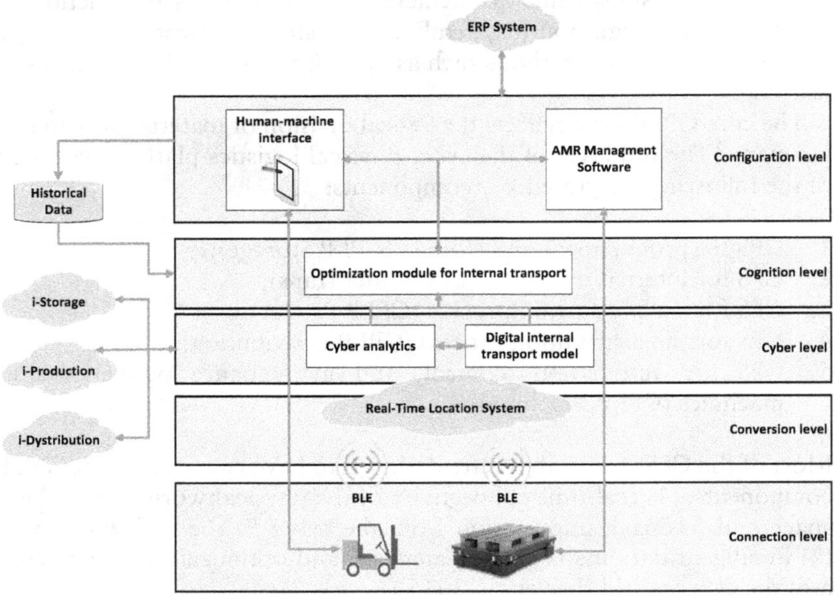

Figure 9.2 Conceptual structure of CPLP storage.

The CPLP-InterTrans subsystem includes two main models: the digital internal transport model and the optimization model for the dynamic planning of routes for the heterogeneous fleet of vehicles. The internal transport system of the printing company includes all the material flows for production, a finished goods flow from production, and a reverse flow (return materials, empty transportation units, etc.). The internal transport fleet combines the MiR1000 automated guided robot vehicle with a traditional forklift (with a human operator). Currently, it is impossible to implement a fully autonomous fleet due to AMR weight constraint. The current locations of the forklifts are gained from a real-time location system (RTLS) that can handle non-homogeneous data. Figure 9.3 presents the architecture of the internal transport within the CPS.

The current localization of the forklifts is broadcast in real time by location sensors using the BLE technology. Next, the raw data is transformed into information at the connection level. At the cyber level, all objects of the CPLP platform exchange information through cyber interfaces. For example, the prediction of a short-term production plan could be taken into account to change the priority of the forklift assignments in the optimization module or the current occupancy of buffer places in the warehouses could influence the optimization constraints. The top

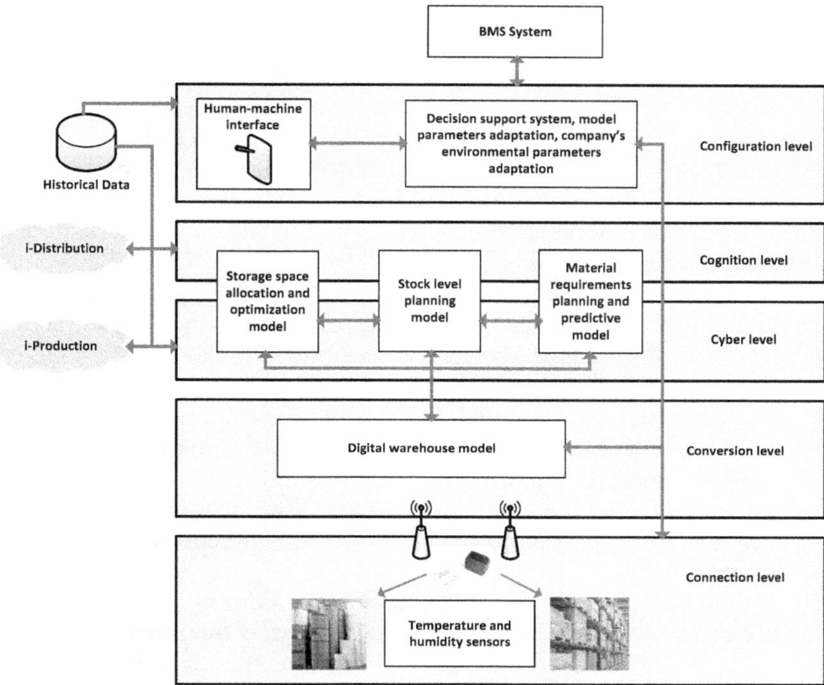

Figure 9.3 Architecture of CPLP InterTrans.

configuration level supervises the correct transmission of the decisions to the lower layer. In this case, the software distributes the transportation orders through the AMR units.

The automation and failure-prediction subsystem (CPLP Machines) consists of the following four main modules:

1. Machine control module—ensuring full visibility of the production processes in real time and warning about problems such as micro-stoppages, excessive vibrations, or other deviations from the standard operating modes of the machines;
2. Machine autosetup module—allows automatic or semi-automatic (via HMI) machine settings for repetitive or similar production tasks;
3. Predictive model for detecting an increased risk of machine failure—using machine-learning models to detect anomalies in data series flowing from both the PLCs and the BMS sensors;
4. Module for planning periodic maintenance—based on historical data and the current condition of the machines as well as the

production plans, it is able to determine the optimal dates for periodic maintenance.

The listed modules use the information provided by the digital models of the production line machines (digital twins) and other production devices like cobots. The models are also responsible for processing and sharing information from the BMS sensors located at the machines. For the machines, data can be read directly from the PLC controllers; however, this is most often done via the SCADA system (the data from cobots is available via dedicated APIs).

The role of CPLP-Production (Figure 9.4) is primarily planning (both on the long- and short-term horizons) as well as tracking the production progress. CPLP-Production consists of the following three main modules:

1. Dynamic planning module—responsible for (a) long-term planning based on the capacity planning and balancing model (i.e., taking long-term contracts and sales forecasts into account) and (b) operational planning (i.e., job sequencing, deadlines, current machine availability, actual production progress, and current production costs).
2. Production control module—tracking the current production progress, possible deviations from the plan, and monitoring the production standards to precisely determine the estimated time of completing production jobs.
3. Short-time material control—verifies the availability of materials; primarily paper, but also entrusted materials necessary for production (inserts and other additions).

Finally, the CPLP-Distribution module optimizes the external logistic process and uses three main models:

1. Transport planning model—including platform transshipments, final products distribution, or semi-finished products transport to subcontractors;
2. Predictive model for flow and demand planning for the distribution of finished goods—its main purpose is to determine the required transport in specific directions;
3. Model for optimizing transport routes—for the vehicles that the company can manage.

In the current business context, it is not possible to use GPS sensors to determine the position of vehicles (primarily for economic and organizational reasons), so no digital model of external transport was built. However, the CPLP-Distribution module can be expanded in the

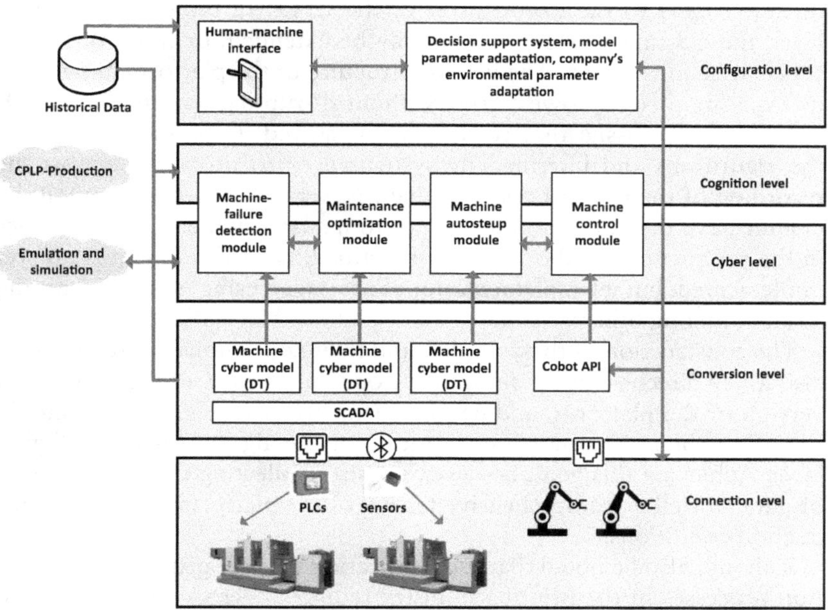

Figure 9.4 Architecture of CPLP machines.

future with such a model, which will allow for the dynamic response of the system, for example, to deviations from the planned routes and delivery dates.

Implementation and Future Extensions

The architecture of the CPLP presented in this chapter is not strictly compatible with most CPSs known from the literature; this is due to the dualism adopted in its design. The platform combines digital models of machines and physical facilities (e.g., forklifts) with models that are built based on data from transactional systems—WMS, TMS, and lower-level systems monitoring the parameters of machines (SCADA) and devices (e.g., RTLS). The basic idea behind the platform was to digitize all key processes in the entire logistics chain, including the elements that currently are not fully monitored (digitally reflected) such as the distribution to end customers. Of course, these elements can be replaced in the future by their digital equivalents (e.g., by introducing a GPS-based distribution fleet-tracking system). The same applies to the development of digital twins and digital images of physical objects. It is also possible to add new physical objects such as new machines, more autonomous vehicles, or new robots to the lowest layer of the CPLP. In the communication layer, the existing SCADA system may be extended, or more BMS sensors can be introduced. The flexible structure of the platform allows one to easily modify the lowest layer without disrupting its functioning. Of course, the key issue that remains to be solved is the modification of the algorithms and intermediate systems used to properly support the recording of the new process variables. Nevertheless, the system will by default have the ability to auto-adapt to many new facilities or changes in the production conditions. This functionality of the system is not yet implemented, but its implementation is envisaged using machine-learning models, among others.

The construction of these models and their continuous improvement is also a key direction in the future development of the CPLP. In the target version of the platform, adding new elements or changing them should not affect the models nor the operation of the algorithms in the digital layer. Achieving this goal, however, requires collecting the right amount of data as well as comprehensive testing of the platform under real production conditions.

It should also be noted that the digitization of the logistics and production processes in the printing industry requires overcoming many additional barriers when compared to production in other industries (e.g., electronic, pharmaceutical). First, the processes may behave erratically depending on the environmental parameters (e.g., elevated temperatures) or material parameters (e.g., poor-quality paper can lead to breaks in a printing process). An additional factor that hinders digitization is the speed of some processes, for example, the web speed on an offset printing machine can reach 15 m/s. This requires fast sensors that will potentially generate large amounts of data; hence, data aggregation is necessary and algorithms processing such aggregated data must be developed accordingly. The possible dustiness in some production areas (paper dusting) will also require additional treatment like the cleaning of sensors. The same concerns any vibrations arising on printing machines. Leveling such disturbances requires proper data processing (e.g., smoothing, removal of peaks) and the adaptation of this process—for example, the algorithm learns which peaks are deviations and which carry important information.

References

Benayoun, R., de Montgolfier, J., Tergny, J., & Larichev, C. (1971). Linear programming with multiple objective functions: Step method (STEM). *Mathematical Programming*, *1*, 366–375.

Boschert, S., & Rosen, R. (2016). Digital twin—the simulation aspect. In *Mechatronic futures* (pp. 59–74). Springer.

Frazzon, E. (2005). Big data applied to cyber-physical logistic systems: Conceptual model and perspectives. *Brazilian Journal of Operations and Production Management*, *12*(2).

Klötzer, C., & Pflaum, A. (2015). *Cyber-Physical Systems (CPS) in supply chain management: A definitional approach*. NOFOMA 2015 Post Conference Proceedings: Towards Sustainable Logistics and Supply Chain Management, Molde, Norway. Nordic Logistics Research Network Publisher.

Kong, X. T. R., Zhong, R. Y., Zhao, Z., Shao, S., Li, M., Lin, P., Chen, Y., Wu, W., Shen, L., & Yu, Y. (2020). Cyber physical ecommerce logistics system: An implementation case in Hong Kong. *Computers & Industrial Engineering*, *139*, 106170. doi:10.1016/j.cie.2019.106170

Lee, E. A., & Seshia, S. A. (2014). Introduction to embedded systems—a cyber-physical approach. *LeeSeshia.org*. Edition 1.5, 2014. http://LeeSeshia.org

Lee, J., Bagheri, B., & Kao, H. A. (2015). A cyber-physical systems architecture for industry 4.0-based manufacturing systems. *Manufacturing Letters*, *3*, 18–23.

Liu, C. H., & Zhang, Y. (Eds.). (2015). *Cyber physical systems architectures, protocols and applications*. CRC Press.

'MiR.' (2020). Available at: https://www.mobile-industrial-robots.com/en/solutions/robots/mir1000/

Monostori, L. (2014). Cyber-physical production systems: Roots, expectations and R&D challenges. *Procedia CIRP*, *17*, 9–13. doi:10.1016/j.procir.2014.03.115

Panetto, H., Iung, B., Ivanov, D., Weichhart, G., & Wang, X. (2019). Challenges for the cyber-physical manufacturing enterprises of the future. *Annual Reviews in Control*, *47*.

Pascual, D. G., Daponte, P., & Kumar, U. (Ed.). (2019). *Handbook of industry 4.0 and SMART systems*. CRC Press.

Pujo, P., & Ounnar, F. (2018). Cyber-physical logistics system for physical internet. In T. Borangiu, D. Trentesaux, A. Thomas, & O. Cardin (Eds.), *Service orientation in holonic and multi-agent manufacturing. Studies in computational intelligence*, *762* (pp. 303–316). Springer. doi:10.1007/978-3-319-73751-5_23

Rudtsch, V., Gausemeier, J., Gesing, J., Mittag, T., & Peter, S. (2014). Pattern-based business model development for cyber-physical production systems. *Procedia CIRP*, *25*, 313–319. doi:10.1016/j.procir.2014.10.044

Seitz, K. F., & Nyhuis, P. (2015). Cyber-physical production systems combined with logistic models—a learning factory concept for an improved production planning and control. *Procedia CIRP*, *32*, 92–97.

Tao, F., Cheng, J., Qi, Q., Zhang, M., Zhang, H., & Sui, F. (2018). Digital twin-driven product design, manufacturing and service with big data. *The International Journal of Advanced Manufacturing Technology*, *94*(9–12), 3563–3576.

Tao, F., Qi, Q. Q., Wang, L., & Nee, A. Y. C. (2019). Digital twins and cyber—physical systems toward smart manufacturing and industry 4.0: Correlation and comparison. *Engineering*, *5*(4), 653–661.

Trappey, A. J. C., Trappey, C. V., Govindarajan, U. H., Sun, J. J., & Chuang, A. C. (2016). A review of technology standards and patent portfolios for enabling cyber-physical systems in advanced manufacturing. *IEEE Access*, *4*, 7356–7382.

Venkata Sundeep, B., & Sree Vardhan, C. (2013). Telematics and its applications in automobile industry. *International Journal of Engineering Trends and Technology*, *4*(4), 554–557.

Xu, L. D., Xu, E. L., & Li, L. (2018). Industry 4.0: State of the art and future trends. *International Journal of Production Research*, *56*, 1–22.

10 Autonomous Vehicles as Part of Industry 4.0 Concept

Mariusz Trela

Introduction

Being an extremely important link in supply chains, transport is also an integral part of the production process to which this new approach is concerned. Additionally, the fact that transport and logistics operators develop and implement new solutions in supply chain management and transport systems (Paprocki, 2016, pp. 186–187) means that transport must be considered as a component of the Fourth Industrial Revolution. Having achieved the largest share of transport work of all transport branches in Poland and Europe,[1] road transport is a particularly important link in the process of implementing the concept of Industry 4.0. At the same time, the context of the definition of this concept first of all includes the revolution taking place in the aspect of the Internet of Things (IoT), whose revolution is believed to be leading this branch to the introduction of autonomous vehicles (its entry into actual operation will mean the complete readiness of the road transport branch to implement the assumptions of Industry 4.0). Optimistic forecasts assume that fully autonomous cars will be available in 2035 (Bimbraw, 2015, p. 196). Hopes for a relatively quick accomplishment of these assumptions are further stimulated by the fact that creating an autonomous commercial vehicle is assumed to be easier than creating a passenger car (Gaffar & Kouchak, 2017, p. 5); it is commercial vehicles that constitute the basis for the revolution in the industry.

Therefore, it seems important to analyze the branch of the economy that road transport is in terms of its current position in the area of developing modern information technologies on the way to achieving the fifth level of automation (i.e., creating a fully autonomous vehicle according to the methodology used by SAE[2]), which is the aim of this article. The identification of possible developmental trends and any related challenges is a desirable additional element of this analysis, the implementation of which is also undertaken in the article.

DOI: 10.4324/9781003186373-10

State of Research in Context of Autonomous Transport

In the field of autonomous transport, a lot of research and analysis has been carried out; however, this has mainly concerned the proposal of improving and testing each driving automation system according to one's own non-standardized criteria. Such research was conducted more than 10 years ago (Buhler & Wegener, 2004) and concerned basic systems in today's cars—parking systems. The current research focuses primarily on systems that take control of a car (Lv et al., 2018; McCall et al., 2019; Mimura et al., 2020; Paul & Chung, 2018; Zaarane et al., 2020; Zein et al., 2018).

Equally, as many analyses have been made in predicting the potential opportunities and threats of autonomous transport (Grush & Niles, 2018; Knight, 2020; Knoefel et al., 2019; Kockelman & Boyles, 2018; Martin, 2019; Meyboom, 2018; Shaheen et al., 2018). Many publications also concern the social acceptance of such vehicles (Bansal et al., 2016; Krueger et al., 2016; Wu et al., 2020; Zmud et al., 2016). On many occasions, the literature has also referred to the estimated time of putting autonomous vehicles into service (Davidson & Spinoulas, 2015; Litman, 2015) or to the costs and benefits associated with purchasing and operating such vehicles (Bansal et al., 2016; Berrada & Leurent, 2017; Chen et al., 2016; Johnston & Walker, 2017; Keeney, 2017).

However, it is extremely difficult to find an attempt to verify the actual level of vehicle automation in the literature; this topic has been undertaken in this article. There are statements regarding the currently available level of vehicle autonomy without verifications of their correctness (Teoh, 2020, p. 145) or the assumptions of vehicle manufacturers (Mordue et al., 2020, p. 176), which seem to be not very reliable considering the marketing aspects. One of the most important reasons for this is that there are no official standards for controlling the level of vehicle automation and meeting an autonomy system's requirements. Several organizations are working on creating standards for such systems (International Telecommunication Union [ITU], Underwriters Laboratories [UL], and British Institute of Standards [BSI Group]); however, no actual standards have been recognized as being official.

In practice, the only permanent reference point that is formally or informally accepted in the entire automotive world (e.g., formal acceptance by the National Highway Traffic Safety Administration [NHTSA] in the United States) in the verification of compliance with the requirements of an autonomous vehicle is the SAE definition, which defines six levels of driving automation.

This path of qualifying vehicles for automation levels seems to be even more justified, as the Automated Vehicle Safety Consortium has developed standards for the fourth and fifth levels of driving automation in accordance with this definition. Given the general acceptance of the SAE definition (Lee & Hess, 2020, p. 86), this provides hope for the relatively short creation time of a formal classification system that enables the assignment of levels of automation based on this definition. Therefore, it was decided to conduct the analysis in this article assuming requirements in accordance with the SAE definition.

Automated Vehicle and Autonomous Vehicle

When creating marketing campaigns in recent years, car manufacturers have placed a great deal of emphasis on the driver-assistance systems used in particular vehicle models. Advertising materials are prepared in such a way that the recipients can very easily get the impression that the systems that are currently mounted in cars take over some of the driver's driving duties (especially if they do not analyze them very carefully). At the same time, car manufacturers are creating a marketing reality in which a modern currently produced car with many driver-assistance systems is identified as an "almost autonomous" vehicle. However, the reality is quite different; it is only in certain precisely defined and specific cases that the most modern and currently available systems are capable of performing a particular maneuver without a driver's intervention—they never relieve the driver of their obligation to maintain constant control over a vehicle. Therefore, the operation of systems that are currently available should be understood not in terms of discharging one from the obligation to drive but only as an absence of the need to perform certain technical operations (but not always, and only under strictly defined conditions). This causes the vision of an autonomous car to move very far away; thus, it is a signal to drivers that they have identical responsibilities and must analyze every traffic situation as carefully as with vehicles manufactured several years earlier. Any doubts in this respect are explained by the theoretical classification of the degree of automation of a vehicle developed by SAE International (Figure 10.1).

The most important thing that becomes evident from this classification is that a Level-5 autonomous car is a vehicle that operates independently under all conditions and in all situations without driver intervention. Vehicles that do not meet this condition are not autonomous vehicles but only automated to a varying extent.

From the point of view of implementing the Industry 4.0 vision in the context of the Internet of Things (IoT) (i.e., in this case, the machine assignment of orders for individual vehicles, the communication between vehicles and customers, and recipients of the communication of vehicles with each other), it is necessary to achieve Level-5 vehicle automation. In specific cases (e.g., the execution of orders on a route or in an area that is precisely defined and properly prepared in terms of infrastructure), it might be sufficient to reach Level 4.

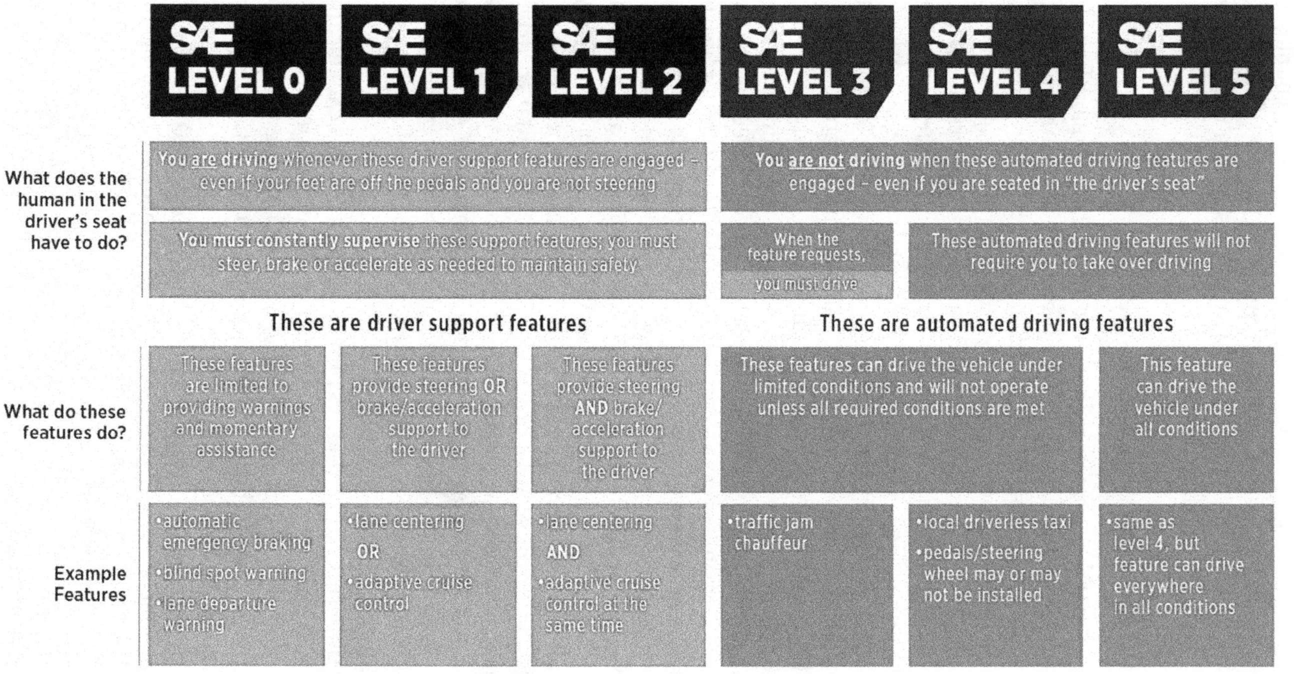

SAE LEVEL 0
SAE LEVEL 1
SAE LEVEL 2
SAE LEVEL 3
SAE LEVEL 4
SAE LEVEL 5
What does the human in the driver's seat have to do?
You are driving whenever these driver support features are engaged – even if your feet are off the pedals and you are not steering
You are not driving when these automated driving features are engaged – even if you are seated in "the driver's seat"
You must constantly supervise these support features; you must steer, brake or accelerate as needed to maintain safety
When the feature requests,
you must drive
These automated driving features will not require you to take over driving
These are driver support features
These are automated driving features
What do these features do?
These features are limited to providing warnings and momentary assistance
These features provide steering OR brake/acceleration support to the driver
These features provide steering AND brake/acceleration support to the driver
These features can drive the vehicle under limited conditions and will not operate unless all required conditions are met
This feature can drive the vehicle under all conditions
Example Features
•automatic emergency braking
•blind spot warning
•lane departure warning
•lane centering
OR
•adaptive cruise control
•lane centering
AND
•adaptive cruise control at the same time
•traffic jam chauffeur
•local driverless taxi
•pedals/steering wheel may or may not be installed
•same as level 4, but feature can drive everywhere in all conditions

Figure 10.1 Levels of automobile automation

Source: © SAE International from SAE J3016™ Taxonomy and Definitions for Terms Related to Driving Automation Systems for On-Road Motor Vehicles (2018-06-05), https://www.sae.org/standards/content/j3016_201806/

Commercial Vehicles in Concept of Industry 4.0

Commercial vehicles are the one element of road transport that can most effectively implement the assumptions of the Industry 4.0 concept. Heavy-duty vehicles occur very often in transport systems, and they often generate very significant costs in the entire logistics process. The benefits of introducing automation in this type of vehicle can be related to the following cost items:

- Personnel costs (drivers);
- Vehicle operating costs;
- Costs of vehicle insurance and damages (related to damage of vehicles and loads);
- Cost of delays.

The potential benefits of using autonomous commercial vehicles could be very important for most entrepreneurs. However, the systems used in passenger cars are pioneers in the automation of driving in road transport. This results from the fact that passenger vehicles are cheaper to test, easier to control in normal situations, and much easier to control in extreme emergency situations when compared with trucks; additionally, any damage caused by errors in vehicle control systems is incomparably smaller. These factors determine that passenger cars are the most technologically advanced in driving automation systems. After a period of testing and refinement, these systems eventually become equipment for commercial vehicles.

Only in commercial vehicles is a system being tested that allows trucks to move in convoys (platooning)—several vehicles following one-by-one at close distances. The assumption of this system is to reduce driver fatigue and reduce fuel consumption by using the aerodynamics of the vehicle ahead. Such a system has been tested by Mercedes-Benz and MAN. In passenger cars, the platooning system has not been tested because it would not be used in practice; from a technological point of view, however, it is based on systems used in passenger cars (active cruise control, lane assistance).

The automation systems used in mass-produced commercial vehicles have less functionality than the analogous systems from passenger cars. Although there are companies in the United States (e.g., Plus.ai, TuSimple, Embark) that have informed the public that their trucks have traveled several thousand kilometers in highway traffic without driver intervention, there is no evidence describing that fact. Moreover, these systems are not the factory equipment of mass-produced vehicles but only modifications, so they should be considered to be prototypes. Similarly, theoretically extremely technologically advanced Tesla and Nikola trucks should be considered to be prototypes. A Volvo Vera that moves autonomously (but only on a certain route between the logistics center and the Goteborg port terminal with a limited speed) cannot be considered to be a mass-produced commercial utility vehicle either.

In the context of the Industry 4.0 concept, however, heavy-duty vehicles have a huge advantage over passenger cars—they are mainly used in intercity traffic on well-developed road infrastructures (motorways), which means that a relatively small number of variable parameters must be correctly recognized and interpreted by the vehicle management system (as opposed to urban traffic). In the case of commercial vehicles, there is even the possibility of the very thorough preparation of a road infrastructure on a strictly defined route. Then, a vehicle that was directed to handle only this connection would have a well-defined set of data that is needed to move. It is possible to repeat a specific section of the route many times, analyze the missing information needed to make the optimal decision by the driving system, and always equip the infrastructure with elements that provide missing information to the vehicle. In this way, it is possible to prepare an automated vehicle to carry out transport tasks on a strictly marked and prepared route. In many cases, this approach may prove to be sufficient to significantly reduce the costs associated with the transport chain. This gives grounds for assuming that commercial vehicles will be able to perform the tasks according to the concept of Industry 4.0 sooner than passenger cars will reach the SAE Level 5 of automation driving.

Methodology of Analysis

In order to achieve the aim of the paper, a matrix has been created (Table 1) containing the currently used solutions for automated driving in the rows and driving cycles (urban driving, driving in congestion or slow traffic, extra-urban driving, driving on expressways and motorways, and parking) in the columns. The cycles are meant to reflect the division of actual driving conditions into individual groups that are characterized by characteristic driving parameters. Each of the systems has been analyzed individually in terms of meeting the assumptions of the particular levels of driving automation contained in the SAE International definition. On this basis, a rating within a range of 0–5 (corresponding to the level of driving automation as defined) has been allocated to each system for each driving cycle. This approach makes it possible to identify those areas (driving cycles) not covered by the particular systems.

Then, the synergy effects resulting from the simultaneous operation of the individual systems were analyzed. This made it possible to determine the theoretical maximum level of driving automation that is currently achievable. At the same time, it is possible to determine whether the maximum level of driving automation thus determined applies to all driving cycles based on the matrix (i.e., the functioning of the vehicle under all road conditions). This approach makes it possible to identify any shortcomings in the driving automation systems and formulate the objectives necessary to achieve the fourth or fifth levels of the driving automation target from the point of view of Industry 4.0.

Although commercial vehicles will be of crucial importance in the Industry 4.0 concept, the paper decides to carry out the analysis on the basis of systems used in passenger cars, as the driving automating technology used in commercial vehicles comes mostly from passenger vehicles (where it is developed and refined and only later applied to commercial vehicles). At the same time, more diverse systems are being used in passenger cars, making it possible to analyze in a broader context the currently available technological solutions that have a chance of being incorporated into commercial vehicles in the future. Therefore, by analyzing the systems used in passenger cars, it is possible to draw conclusions for commercial vehicles that relate not only to the present but also to the near future.

Overview of Currently Applied Solutions and Their Classifications

All major companies selling their vehicles on the European and American markets use systems to automate certain driving activities in their vehicles. The number of systems used and their sophistication depend on the price of the vehicle, but all relevant systems can be divided into the following groups:

- Active cruise control (automatic braking function is part of this system);
- Lane assist;
- Blind spot assist;
- Recognition of traffic signs;
- Lane change system;
- Automatic parking;
- Cross-traffic detection;
- Pedestrian detection.

Adaptive Cruise Control (ACC) (Automatic Braking Function Is Part of This System)

ACC vehicles detect information about vehicle(s) directly ahead via vehicle-mounted sensors and make longitudinal reactions automatically (Tu et al., 2019, p. 177). This means that the system is not only able to maintain a set speed but (above all) react to a change in the speed of the vehicle ahead and automatically change the speed or even stop the vehicle where ACC is installed. One limitation of such systems is detecting changes in direction. Such a system does not allow one to change lanes or follow a lane during a turn. The system can only function under the constant control of the driver.

When analyzing the functions and limitations of this system, it should be assumed that it fulfills the assumptions of Level-1 driving automation. It is responsible for accelerating, braking, and maintaining speed but does not control the direction of travel.

Lane Assist

Such systems focus on maintaining the set direction of travel of its vehicle and are intended to prevent uncontrolled lane changes. They track the position of the vehicle in relation to the lines separating the road as well as the roadside and warn the driver if an unintentional lane change is detected. Lane departure warnings may include haptic (e.g., steering wheel vibration) as well as audible and/or visual elements (Eichelberger & McCartt, 2016, p. 68). More advanced systems are able to correct the vehicle's path if they detect a lane change maneuver without the turn signal being activated and are able to keep the vehicle in one lane without driver intervention.

Considering the functionality of such systems, two main groups can be distinguished:

1. Systems that detect inadvertent lane departure and alert the driver but are not capable of interfering with their vehicle's path.
2. Systems that are capable of impacting their vehicle's path without driver intervention.

Systems belonging to the first group should be classified as Level 0 with regard to driving automation. Systems in the second group meet the assumptions that concern Level-1 driving automation.

The simultaneous use of adaptive cruise control (ACC) and Category-2 lane assistance in a vehicle makes it possible to classify the vehicle as Level 2-automated because, when combined, these systems make it possible to control the vehicle's path as well as its speed.

Blind Spot Monitoring System

This system is intended to inform the driver of a moving vehicle that is within close proximity, which may be invisible due to the driver's so-called "blind spot" in the mirrors. This system only has an information function; as such, it is not an automation system (Level 0). However, this system is an indispensable complement to the systems that allow automatic changes of direction of a vehicle (lane change assistance).

The coexistence of these two systems does not mean that a vehicle can be classified as meeting the requirements of at least Level-1 automation, as the driver must be involved in any lane change (activating the turn signal). If a change of lanes were possible with a computer activating the turn signal, then it would be possible to classify such a vehicle as meeting the requirements of Level-1 driving automation (where ACC systems are not installed) or Level-2 driving automation (with the additional coexistence with ACC systems).

Traffic Sign Recognition (TSR) Systems

Traffic sign recognition systems can be used to inform a driver about any currently applicable traffic signs (e.g., the current speed limit). They can also be a module that extends the functionality of adaptive cruise control, adjusting vehicle speed not only to match the vehicle(s) ahead but also to the current speed limit. Regardless of the technological sophistication of the system, however, it can only be a source of information for either the driver or another in-vehicle system. This feature qualifies the system to Level 0 in the context of the SAE definition.

Although it is often claimed that automatic traffic sign detection and recognition play crucial roles in several expert systems such as driver assistance and autonomous driving systems (Berkaya et al., 2016, p. 67), it should be noted that adding this functionality to an ACC system will not result in any change in the vehicle classification according to the SAE definition. This is due to the fact that the additional information that can be obtained in this way is greatly inadequate to enable a vehicle to attain Level-3 driving automation. However, this does not change the fact that a TSR system needs to be installed in the vehicle to reach this level.

Lane Change System

This system allows a vehicle to change lanes, but the driver must be involved in this activity—the turn signal must be activated, but the rest will be done by the vehicle itself. The operation of this system allows for a change of direction and, in combination with Adaptive Cruise Control (ACC), causes the car to accelerate and brake automatically; however, it will not allow the vehicle to meet the requirements for at least Level 1 of the SAE definition, as it is necessary for the driver to activate the turn signal throughout the process. In theory, however, there is a system that is able to change lanes completely without the driver's involvement; however, in vehicles across Europe, this functionality has been limited by the manufacturer (Tesla) to such an extent that the system also does not meet the requirements for Level 1 of the SAE definition (except when driving on multi-lane expressways or motorways).

Automatic Parking

Computer systems are currently capable of parking a vehicle—both in perpendicular and parallel parking situations. As a standard, these systems scan a space, check that it is large enough to park the vehicle, and take control of the steering wheel (leaving the driver in control of the pedals). More advanced systems are able to perform the parking themselves, leaving the driver to only decide where to park. This makes it possible to classify these most advanced systems to Level-3 driving automation (the driver must observe the correctness of the maneuver at all times) but with the express stipulation that this level only applies to a very specific case—one specific parking case. However, not all parking spaces where the driver is able to park the vehicle will be positively qualified by the system for this maneuver. In addition, the driver must always be ready to take over control of the vehicle. This makes it impossible to accept this solution as Level-4 driving automation. Nevertheless, granting this system such a high level of automation does not significantly bring the vehicles closer to reaching SAE Level 3, as parking is a maneuver that generally takes place in a very limited area that has previously been studied and accepted by the computer system during the scanning process. In addition, this cycle requires controlling a relatively small number of variable parameters as compared to driving cycles (especially the urban driving cycle). Combining this system with other driving automation systems will in no way increase the overall assessment of the level of driving automation for cars according to SAE, as only Level 3 is achievable in the parking cycle.

Cross-Traffic Detection

This is a system that detects a vehicle approaching from the side (e.g., when reversing from a parking space) and warns the driver; if the driver does not take the appropriate action to avoid a collision, the system will stop the vehicle. This system is mainly used in urban traffic during maneuvers that are related to parking and merging with traffic. The system therefore meets the requirements for Level 1 of the SAE definition for selected driving cycles. Joint operation of this system with other driving automation systems will not increase the level of automation of the vehicle from the SAE definition's point of view.

Pedestrian Detection

This system detects pedestrians in or approaching a vehicle's path. If the system detects a risk of collision with a pedestrian, it will warn the driver and automatically stop the vehicle if no action is taken. The system can also detect animals in a similar way. The importance of this system for driving automation is very similar to that of the cross-traffic detection system with one difference—pedestrian and animal detection can theoretically be applied in any driving cycle. However, this does not change the fact that the system meets the requirements for Level-1 driving automation; its coexistence with other systems will not increase the level of vehicle automation in any way in the context of the SAE definition. At the same time, the presence of both systems (cross-traffic detection and pedestrian detection) is necessary to achieve Level-4 driving automation.

Results of Analysis

The effectiveness of the examined systems has not been verified by any standardized tests (standards do not exist) to ensure an objective assessment; it is therefore impossible to conclusively determine whether and to what extent these systems work properly. Similarly, it is impossible to create a classification of vehicle models that defines their technological advancement.

It is therefore assumed that these systems work as intended by the manufacturer of the vehicle in which they are installed. For some systems, the available sources indicate that tests have been carried out to assess their effectiveness but that their selectivity and the lack of standardized test procedures make it impossible to draw general conclusions on this basis.

On the basis of the analysis, a matrix showing the current state of the automation of road transport vehicles was created (as shown in Table 10.1).

In Table 10.1, the value of "3" for the level of vehicle automation resulting from the synergy effect for the "congestion and slow traffic" and "parking" cycles is notable. While the Level-3 automation was justified for the "parking" cycle in the analysis of the relevant systems, this may raise some doubts with regard to the "congestion . . ." cycle. The paper assumes that the system works as foreseen by the vehicle manufacturer and that there are not sufficient grounds (standardized performance tests) to challenge the manufacturer's information. Audi is the only manufacturer to officially offer its vehicle (Audi A8) as compliant with Level-3 automation according to the SAE definition. Since this function works with speeds of up to 60 km/h and is intended for slow traffic, it was decided to assume that the Level-3 automation of road vehicles is currently only possible under the conditions of congestion and slow traffic (excluding parking). In practice, even this cycle is limited to congestion on motorways and expressways (excluding city driving) because the Audi system only operates on roads with at least two lanes in each direction that are separated by a barrier or median strip and when there are no pedestrian crossings or traffic lights in the field of view (Denton, 2020, pp. 124–128). The computer system functions that are offered by

Table 10.1 Automation level of motor vehicles as defined by SAE International

	city	*congestion and slow traffic*	*out-of-city traffic*	*expressway or motorway*	*Parking*
Adaptive cruise control (automatic braking function is part of this system)	1	1	1	1	-
Lane assist	1	1	1	1	-
Blind spot assist	0	0	0	0	-
Traffic sign recognition	0	0	0	0	-
Lane change system	0	0	0	1	-
Automatic parking	-	-	-	-	3
Cross-traffic detection	0	0	0	0	1
Pedestrian detection	1	1	1	1	1
Automation level for vehicle resulting from synergy effect	2	2/3	2	2	3

Source: own study

other manufacturers theoretically provide very similar functionalities, with the difference being that manufacturers (even Tesla) officially define their systems as meeting the requirements for the Level-2 automation of vehicles. From the technical point of view, it therefore seems justified to take the view that vehicles in this particular cycle (with some additional restrictions) are already set for Level-3 automation according to the SAE definition. If, on the other hand, the operation of driving assistance systems is interpreted in the context of the impossibility of demonstrating Level-3 driving automation under all conditions of the "congestion and slow traffic" cycle (this function is practically unavailable in urban congestion, which is the most common case), then the level of vehicle automation in that cycle should be assessed as Level 2—hence, the 2/3 rating awarded.

At the same time, it should be noted in the matrix that the "expressway or motorway" cycle is also very close to Level 3, which is due to the fact that it is only in this cycle that the functionality of automatic lane changing is operative. Based on the results of the analysis, it can therefore be stated that it is the most difficult for manufacturers to achieve higher levels of automation in urban traffic, where traffic is very busy with changeable vehicle environments, and in the extra-urban cycle where lanes with opposite traffic directions are not separated and/or where there is only one lane in each direction.

Conclusions

The analysis showed that mass-produced vehicles can fully meet Level-2 and (only in some situations) Level-3 automation driving, according to the SAE definition.

The concept of Industry 4.0 can only be fully accomplished when autonomous vehicles that meet the requirements for at least Level 4 as defined by SAE are available for road transport. For this to be possible, it is necessary to improve all systems in each driving cycle. Vehicle manufacturers should therefore develop their products in the following directions:

- The creation and introduction of technical devices adapted to analyzing all of the necessary parameters coming from the inside of the vehicle and its surroundings as well as enabling the processing of such amounts of data (hardware);
- The creation and introduction of software that is capable of analyzing, drawing appropriate conclusions, formulating, and transmitting appropriate signals to those devices responsible for driving the vehicle (software).

Manufacturers should develop software in vehicles to such a level that a vehicle control system would be able to repetitively and faultlessly accomplish the following:

- Automatic lane changing in each driving cycle;
- Detecting vehicles moving in directions other than that of the computer-controlled vehicle;
- Keeping the vehicle in lanes where large steering angles are required;
- Turning when changing directions (e.g., at a crossing);
- Correctly responding to traffic lights and pedestrian crossings.

It seems that, in the near future, it will be possible to implement the basic assumptions of the Industry 4.0 concept in the field of heavy-duty transport. For this purpose, it would be advisable for the user of autonomous vehicles to realize the following actions:

- Defining a permanent route for automated transport;
- Testing different models of vehicles to determine which of them best perform transport tasks on a defined route in "standard" automated driving settings;
- Selecting a specific model of the vehicle and conducting tests on a defined route, focusing on identifying any missing information that precludes the automated driving system making optimal decisions;

- Equipping the route with the appropriate infrastructure that enables the vehicle control system to obtain the greatest amount of valuable information;
- Optimally configuring the vehicle control system for the defined route.

The above sequence of actions requires the cooperation of the fleet user with the vehicle manufacturer as well as the road administrator. It also requires the involvement of legislators to assure that some technical solutions may be allowed.

Road transport is therefore on the right track toward the "Industry 4.0" concept, but the current level of technology used in mass-produced vehicles does not allow this as of yet.

Notes

1. Based on EUROSTAT Statistics
2. SAE International—originally the Society of Automotive Engineers—is a professional organization established in 1905 that brings engineers together when dealing with, inter alia, the automotive industry. This organization standardizes equipment, such as cars and trucks.

References

Bansal, P., Kockelman, K. M., & Singh, A. (2016). Assessing public opinions of and interest in new vehicle technologies: An Austin perspective. *Transportation Research Part C: Emerging Technologies*, *67*, 1–14.

Berrada, J., & Leurent, F. (2017). Modeling transportation systems involving autonomous vehicles: State of the art. *Transportation Research Procedia*, *27*, 215–221.

Berkaya, S. K., Gunduz, H., Ozsen, O., Akinlar, C., & Gunal, S. (2016). On circular traffic sign detection and recognition. *Expert Systems with Applications*, *48*, 67–75.

Bimbraw, K. (2015). *Autonomous cars: Past, present and future a review of the developments in the last century, the present scenario and the expected future of autonomous vehicle technology*. Proceedings of the 12th International Conference on Informatics in Control, Automation and Robotics (ICINCO), Colmar, pp. 191–198.

Buhler, O., & Wegener, J. (2004). *Automatic testing of an autonomous parking system using evolutionary computation*. SAE Technical Papers. doi:10.4271/2004-01-0459.

Chen, D., Kockelman, K. M., & Hanna, J. P. (2016). *Operations of a shared, autonomous, electric vehicle fleet: Implications of vehicle & charging infrastructure decisions*. Proceeding of the 95th Annual Meeting of Transportation Research Board.

Davidson, P., & Spinoulas, A. (2015). *Autonomous vehicles: What could this mean for the future of transport?* AITPM National Conference, Brisbane, Queensland, Australia.

Denton, T. (2020). *Automated driving and driver assistance systems*. Institute of the Motor Industry IMI and Routledge.

Eichelberger, A. H., & McCartt, A. T. (2016). Toyota drivers' experiences with dynamic radar cruise control, pre-collision system and lane-keeping assist. *Journal of Safety Research*, *56*, 67–73. doi:10.1016/j.jsr.2015.12.002

Gaffar, A., & Kouchak, S. M. (2017). *Determinism in future cars: Why autonomous trucks are easier to design*. Proceedings of the IEEE Advanced and Trusted Computing (ATC 2017), San Francisco, August 4–8. doi:10.1109/UIC-ATC.2017.8397598

Grush, B., & Niles, J. (2018). *The end of driving: Transportation systems and public policy planning for autonomous vehicles*. Elsevier (www.elsevier.com/books/the-end-of-driving/niles/978-0-12-815451-9).

Johnston, C., & Walker, J. (2017). *Peak car ownership: The market opportunity for electric automated mobility services*. Rocky Mountain Institute (www.rmi.org), at http://bit.ly/2rhJRNi

Keeney, T. (2017). *Mobility-as-a-service: Why self-driving cars could change everything*. ARC Investment Research (http://research.ark-invest.com), http://bit.ly/2xz6PNV

Knight, W. (2020). Snow and ice pose a vexing obstacle for self-driving cars. *Wired Magazine*, www.wired.com/story/snow-ice-pose-vexing-obstacle-self-driving-cars

Knoefel, F., Wallace, B., Goubran, R., Sabra, I., & Marshall, S. (2019). Semi-autonomous vehicles as a cognitive assistive device for older adults. *Geriatrics Basel, Switzerland*, *4*(4), 63. doi:10.3390/geriatrics4040063

Kockelman, K. M., & Boyles, S. D. (2018). *Smart transport for cities & nations: The rise of self-driving & connected vehicles*. The University of Texas at Austin, www.caee.utexas.edu/prof/kockelman/public_html/CAV_Book2018.pdf

Krueger, R., Rashidi, T., & Rose, J. M. (2016). *Adoption of shared autonomous vehicles – A hybrid choice modeling approach based on a stated choice survey*. Transportation Research Board 95th Annual Meeting.

Lee, D., & Hess, D. J. (2020). Regulations for on-road testing of connected and automated vehicles: Assessing the potential for global safety harmonization. *Transportation Research Part A: Policy and Practice*, *136*, 85–98. http://doi.org/10.1016/j.tra.2020.03.026

Litman, T. (2015). *Autonomous vehicle implementation predictions: Implications for transport planning*. Victoria Transport Policy Institute.

Lv, C., Cao, D., Zhao, Y., Auger, D. J., Sullman, M., Wang, H., Dutka, L. M., Skrypchuk, L., & Mouzakitis, A. (2018). Analysis of autopilot disengagements occurring during autonomous vehicle testing. *IEEE/CAA Journal of Automatica Sinica*, *5*(1), 58–68.

Martin, G. T. (2019). *Sustainability prospects for autonomous vehicles environmental, social, and urban*. Routledge.

McCall, R., McGee, F., Mirnig, A., Meschtscherjakov, A., Louveton, N., Engel, T., & Tscheligi, M. (2019). A taxonomy of autonomous vehicle handover situations. *Transportation Research Part A: Policy and Practice*, *124*, 507–522. https://doi.org/10.1016/j.tra.2018.05.005

Meyboom, A. L. (2018). *Driverless urban futures a speculative atlas for autonomous vehicles*. Routledge.

Mimura, Y., Ando, R., Higuchi, K., & Yang, J. (2020). Recognition on trigger condition of autonomous emergency braking system. *Journal of Safety Research*, 72, 239–247. https://doi.org/10.1016/j.jsr.2019.12.018

Mordue, G., Yeung, A., & Wu, F. (2020). The looming challenges of regulating high level autonomous vehicle. *Transportation Research Part A: Policy and Practice*, *132*, 174–187. http://doi.org/10.1016/j.tra.2019.11.007

Paprocki, W. (2016). How transport and logistics operators can implement the solutions of "industry4.0". In M. Suchanek (Ed.), *Sustainable transport development, innovation and technology* (pp. 185–196). Springer.

Paul, N., & Chung, C. (2018). Application of HDR algorithms to solve direct sunlight problems when autonomous vehicles using machine vision systems are driving into sun. *Computers in Industry*, *98*, 192–196. https://doi.org/10.1016/j.compind.2018.03.011

Shaheen, S., Totte, H., & Stocker, A. (2018). *Future of mobility*. White Paper. ITS Berkeley, https://escholarship.org/uc/item/68g2h1qv.

Teoh, E. R. (2020). What's in a name? Drivers' perceptions of the use of five SAE Level 2 driving automation systems. *Journal of Safety Research*, *72*, 145–151. http://doi.org/10.1016/j.jsr.2019.11.005

Tu, Y., Wang, W., Li, Y., Xu, C., Xu, T., & Li, X. (2019). Longitudinal safety impacts of cooperative adaptive cruise control vehicle's degradation. *Journal of Safety Research*, *69*, 177–192. doi:10.1016/j.jsr.2019.03.002.

Wronka A., & Szczepka W. (2019). *Znaczenie norm w rewolucji 4.0*. Proceedings of the "Przemysł 4.0" Conference, Polski Komitet Normalizacyjny, Warszawa, May 22.

Wu, J., Liao, H., & Wang, J. W. (2020). Analysis of consumer attitudes towards autonomous, connected, and electric vehicles: A survey in China. *Research in Transportation Economics*, *80*. http://doi.org/10.1016/j.retrec.2020.100828

Zaarane, A., Slimani, I., Al Okaishi, W., Atouf, I., & Hamdoun, A. (2020). Distance measurement system for autonomous vehicles using stereo camera. *Array*, *5*. http://doi.org/10.1016/j.array.2020.100016

Zein, Y., Darwiche, M., & Mokhiamar, O. (2018). GPS tracking system for autonomous vehicles. *Alexandria Engineering Journal*, *57*(4), 3127–3137. http://doi.org/10.1016/j.aej.2017.12.002

Zmud, J., Sener, I. N., & Wagner, J. (2016). *Consumer acceptance and travel behavior – Impacts of automated vehicles*. Texas A&M Transportation Institute. Technical Report.

11 Modern Marketing for Customized Products Under Conditions of Fourth Industrial Revolution

Sandra Grabowska and Sebastian Saniuk

Introduction

For the first time in the history of humanity, technology allows to influence the volume and quality of production as well as the functionality of products by connecting more and more numerous industries with society (Barry et al., 2011; Piccarozzi et al., 2018). Due to the use of advanced ICT technologies, it is possible to more accurately adapt production to customer expectations while maintaining low costs, high quality, and efficiency (Müller et al., 2018). New business models and technologies such as artificial intelligence or additive manufacturing accelerate industry transformation processes, changing current business methods and market structures. All of these artifacts pose new challenges for marketing, which must adapt to the digital architecture of the world (Korena et al., 2015).

On the one hand, the Fourth Industrial Revolution moves production to sophisticated networks of enterprises equipped with intelligent devices, machines, and means of transport—all communicating with each other using new technologies such as cloud computing, big data, and the Internet of Things (IoT) (Donnelly et al., 2015). On the other hand, a world dominated by modern technology creates a wide variety of customer expectations and needs, which creates new challenges for modern manufacturing companies (Azmi et al., 2018). It also means new challenges for the marketing strategy (Marketing 4.0); the greatest of these include the fight for customer attention and the promotion of products that are not physically available yet (as they are to be maximally individualized). Customization affects marketing, forcing the creation of modern marketing. Marketing 4.0 is an evolution of traditional marketing; it is its next stage. The environment, production possibilities, customer expectations, products, and communication channels change; this is followed by marketing (Jara et al., 2012).

Marketing is rapidly transforming under the influence of the global Internet network and ICT solutions. It is evolving toward Marketing 4.0 (the so-called MarTech—technologized marketing), becoming the key

DOI: 10.4324/9781003186373-11

area of an organization's activity, determining its development and openness to innovation, and affecting a comprehensive change in the functioning of customer need-oriented enterprises (Jiménez-Zarco et al., 2017). The high standards and expectations of the consumer regarding personalized production translate as the need to conduct research in the field of marketing communication with the client (aimed at determining his needs and preferences, the organization of direct cooperation between a company and the consumer, and the sales organization sphere). In this chapter, the key challenges of Marketing 4.0 are identified in the context of developing the customization of production under the conditions of the Fourth Industrial Revolution.

Modern Marketing Challenges

Customization as a Key Factor Influencing Market Expectations of Consumers

The use of AI in interfaces for designing new products or improving existing ones will allow companies to better meet consumer expectations and increasing customer participation in personalized production (Fogliattoa et al., 2012; Yang et al., 2018). The previously developed mass customization paradigm means the production of goods and services for the needs of a large market, tailored to the preferences of individual customers at prices close to the prices of mass products and, therefore, devoid of the individualization of value (Jiménez-Zarco et al., 2017; Ranjan & Read, 2016). Mass customization does not always mean designing or modifying a product according to the needs of the consumer. Offering a wide range of products and communicating with customers in a personalized way (as a result of which customers receive products that meet their needs) is also included in mass customization. An example of such a strategy is Amazon.com. There are four types of customization (Grabowska et al., 2020):

- Collaborative (based on cooperation)—assumes the active participation of both parties (in particular, involving the company's dialogue with the client, enabling him to know his needs and expectations) as well as activities aimed at providing the client with a composition of value tailored to his needs;
- Adaptive—involves providing customers with the same product that customers can then configure;
- Cosmetic—a superficial adaptation of the product to customer needs;
- Transparent—made by the company without much customer involvement, sometimes even without their knowledge. This type of customization is used when the client's needs are known on the basis of existing interactions or when they are easily predictable.

Often, a set of instruments for examining consumer preferences and expectations regarding the innovative features of a newly designed product is specialized software that enables a large number of customers to propose unique solutions (variants) for a new product. This approach was used on the Fiat 500 product web page in order to customize this model. This application allowed Fiat to know the expectations of customers regarding the equipment of the car long before assembling the first model. Fiat received more than 160,000 model configurations proposed by customers, which were then evaluated and commented on by other customers.

Virtual product tests that activate consumers and are a great source of information for the producer are also becoming increasingly popular. This approach consists of passing various product variants on to potential consumers in the form of virtual prototypes. An example is the footwear manufacturer Adidas, which replaced the physical prototypes of new footwear models with virtual ones, significantly reducing costs and shortening the time to launch a product on the market.

The need to offer individualized products requires the constant monitoring of customer behavior; this assumes the continuous acquisition and analysis of data on consumer preferences and behavior. This method is used by Pandora.com, Spotify, and YouTube to listen to music. Customization is based on creating a playlist that is based on a set of songs that are similar to those currently being listened to. During the broadcasting of songs, customers can evaluate them; on this basis, a recipient (consumer) profile and customization algorithm are developed.

The process of consumer choice is always the final area of mass customization. The fundamental competence in this area is to support the consumer in identifying his needs while reducing the complexity and nuisance of the choices made. Consumer support is carried out using assortment matching, fast cycle, trial-and-error learning, and embedded configuration. The trial-and-error method is an approach that allows customers to easily build their own models, test them, and then modify them (Choo et al., 2013).

Assortment matching is a process that uses software that automatically creates the recommended product configuration for the customer by comparing the needs of the consumer with the available configuration options. The range is adjusted by clothing manufacturer Zafu, which indicates products tailored to specific types of figures on its website. The company thus customizes by providing personalized recommendations and offering standard products (Chui et al., 2010).

On the other hand, the built-in configuration consists of equipping products with the ability to automatically adapt to the needs of consumers. From the manufacturer's point of view, they are standard products; however, they provide their clients with customized values by automatically adapting. An example of a built-in configuration is the Adidas company and its sports shoes that adjust the hardness of its soles to walking or running conditions.

Coca-Cola was one of the first campaigns to take advantage of consumers' natural needs in terms of individuality and a sense of uniqueness. Its success is undoubtedly a sign of modern times and a perfect market sense. An example is an effective advertising campaign consisting of placing consumer names on the labels of bottles and cans with drinks. Young people willingly bought and advertised the drink with their name on blogs and social networks. Similarly, this applied to humorous terms such as Manager or Friend.

Many sports and cosmetics companies offer personalization services. L'oréal offers personalization at stationary points of sale where the product can be made on site. This applies to the Lancome company (among others), which offers "private foundation," which is similar to the products of the South American Etiude Dom brand (Amoer Pacific Group). German brand Schwarzkopf (Henkel) is preparing to test its "salon-lab" project in hairdressing salons in Western Europe and Japan, where a service of personalized shampoos adapted to the hair of the consumer will be offered.

Marketing—New Trends

Technical progress is changing B2B relationships, eliminating traditional competitive strategies and fostering the emergence of increasingly complex business models that need innovative marketing efforts to accurately reach their clients (Berger, 2016). Following the changing business architecture, the expectations of customers and consumers who are looking for tailor-made products are changing. In addition, the Fourth Industrial Revolution is changing the way that products reach the market. There is an absolute transition from the push model to the pull model, where the final recipients are more closely associated with the producers (Sashi, 2012).

To win, you need to pay attention to personalization; this is the main conclusion from the "What's Hot in Digital Commerce in 2017" report developed by Gartner. Its authors clearly emphasize that the user-experience area is currently the most important battlefield in trade by far. Understanding the individual preferences of users (and, thus, creating their psychographic profiles) helps not only in the selection of specific products or brands but also in the ways of communication to a large extent (Dunn, 2015).

It turns out that we talk a lot about our preferences ourselves, leaving traces of our activity on social networking sites. According to researchers from the University of Cambridge Psychometry Center (who studied the activity of 58,000 Americans), the usual "likes" that we leave on Facebook allow us to easily determine our gender, age, race, and even sexual preferences or relationship statuses. By acquiring such information, companies will be able to tempt their customers with products that will suit their individual preferences. This data is a key tool for Marketing 4.0 (Gordon, 2013).

Marketing 4.0 is understood as moving toward participation and the creation of value in a collaborative way where the customer is able to check, confirm, and be more aware of the reality of a brand. For this reason, this fidelity between reality and promises will build a brand's reputation (Longo et al., 2017).

Having clean and centralized data that can answer business questions at an appropriate level of detail and introducing data integrity and collecting it from all critical places (so-called data silos) responsible for the business process in a company allow one to make better business decisions than those based on so-called managerial intuition. In the context of Marketing 4.0, the beginning of strategy integration is to unify all data from all sources. This applies in particular not only to incoming traffic on the website but also to offline data. Spotify receives 600 GB of information about listeners each day. Netflix constantly observes the behavior of more than 100 million users (Imran et al., 2018; Want et al., 2015).

Perhaps the most important transformation of the last decade (affecting countless areas of our lives) is the dynamic penetration of mobile devices. On the one hand, it affects the blurring of boundaries between the online and offline worlds, as a mobile device allows one to be in contact with the Internet at all times (no one is surprised when a customer is shopping while waiting at the airport). Because customers have mobile devices with them at all times, they assist us and support us in an increasing number of daily activities. As a result, they produce more and more data that accurately describes the lives of a single customer; this data is extremely valuable from a marketing point of view. This trend leads to a change in the balance of power in the market—power is transferred to customers who are connected to each other in a network. In online networks, social networking sites have created new ways of interpersonal interaction that allow one to build relationships without geographical nor demographic barriers (Rospigliosi & Greener, 2015). Based on creating shared value with the consumer, we rely on digital marketing, that is, advertising activities for brands on the web (Lies, 2017).

To carry out a global expansion, you need to be extremely flexible and innovative and enter the online market. Companies cannot perceive customers as a regular target market. A modern customer is looking for product/brand reviews on social media, asking questions on various Internet portals and not believing traditional marketing messages (Maresova et al., 2018).

When designing Marketing 4.0 activities, the needs of specific customer groups should be taken into account. According to Philip Kotler, three influential digital subcultures can be distinguished among modern consumers; these are as follows (Longo et al., 2017):

- Young people—the most influential consumer in the digital network segment. They are the first users of new products and technologies; they are trendsetters; they are a group of consumers who clearly change the rules of the game on the market;
- Women—information collectors and holistic consumers who are very often household managers, financial directors, purchasing managers, and family resources managers. Women spend more money on one-off purchases than men. In addition, women are more loyal than men. When asked about regular customers, 32% of online stores said that they were primarily women; only 8% indicated men (survey conducted for the Ceneo.pl website);
- Network Citizens—social activists, very actively making contacts and communicating with others on the network, expressly feeding the network with online content.

Marketing 4.0 is the era of consumer-focused marketing and the B2H (Business to Human) approach. Focusing on a specific consumer and his needs "here and now" allows a company to improve the customer's experience and his perception of the brand and its products. Enterprises that focus on their customers are far more profitable than those that do not (Suomala, 2002).

Research Methods

The main goals of the research were to determine the current needs of customers as related to the products offered as well as to assess the level of product personalization required by consumers. To achieve this goal, a critical literature analysis and survey research methods were used. Selected results of the research conducted by the authors in 2019 have been presented. As a research tool, a questionnaire consisting of closed questions was used. The research was carried out using the CAWI (standardized Computer-Assisted Web Interview) method. The study was conducted on a group of 504 potential customers and consumers representative of the Polish market.

Most of the respondents represented large and medium-sized cities (59%). It is worth noting that the majority of the respondents assessed their financial situation as good (64.1%) or sufficient (23.24%). Around 13.9% of the respondents declared a very good financial (material) situation. Only 1.8% of the respondents declared a poor financial situation. Selected results of the selection of the respondents are presented in Table 11.1.

Table 11.1 Gender, age, place of residence, and subjective assessment of the financial situation of respondents

Age	*Sex*		*Place of residence*				*Financial situation*			
	F	*M*	*village*	*small town*	*medium-sized city*	*big city*	*very good*	*good*	*not bad*	*Bad*
below 18	16	34	3	3	26	18	13	26	10	1
19–25	122	118	56	50	66	68	23	164	49	4
26–35	30	34	10	11	22	21	10	37	14	3
36–45	34	30	14	15	21	14	12	43	9	0
46–55	19	28	11	9	14	13	4	33	9	1
56–67	15	16	12	5	5	9	6	18	7	0
over 67	4	4	4	2	1	1	2	2	4	0
totals	240	264	110	95	155	144	70	323	102	9

Source: own study

Most respondents are a “new type of customer.” According to Philip Kotler (2016), he is a young person from the middle class living in the city with high mobility and continuous access to the Internet. Most belong to the middle class, hence receiving a satisfactory income that can be spent on goods and services. Starting from a lower socio-economic level, they aspire to higher goals and desire more sophisticated experiences. The pace of their lives is very fast and changeable; they must have everything immediately; they value personalized products and services; they are an attractive target market for marketers.

Among other things, the study looked at the answers to the following questions:

- What are the expectations of consumers regarding personalized production?
- What convinces a customer that a product is attractive?
- How do customers perceive their commitment to the process of creating personalized products?

Results

Surveys on a selected group of potential consumers showed the expectations of modern customers. More than half of the respondents declared an interest in personalized products. The largest group of respondents was interested in personalizing clothing and footwear (62.5% of the respondents), electronic devices (39.5%), ordering personalized dishes in restaurants (37.2%), various personalized types of accessories (31.2%), jewelry (24.9%), and house and garden equipment (including furniture) (23.7%).

Only 21% of the respondents were willing to pay more for personalized products and 47% of the respondents make decisions dependent on the level of the price difference between the standard and personalized product and the type of product.

Our studies show a great interest in personalized products created in various customization strategies. As a reason for purchasing personalized products, respondents most often indicate the uniqueness of a product, emphasizing the impact of its final shape/appearance, greater satisfaction, and comfort of use. As many as 55.5% of the respondents believe that personalized products are unique.

The respondents were asked about which variant of customization they would expect most for specific product groups, such as household appliances, cars and motorization, electronics, clothing and footwear, home and garden equipment, jewelry, beauty/cosmetics, toys, food services, everyday products, and software. They could choose between the following variants (Wortmann & Flüchter, 2015):

- Pure customization—a customer influences the entire production process even at the design stage;
- Tailored customization—a generic prototype is presented to a customer and then tailored according to the customer's individual needs;
- Customized standardization—the assembly is customized. A customer can select from a list of options that are made from standard components;
- Pure standardization—no customization is made by the manufacturer or distributor.

The choice of pure customization was most often indicated by respondents in the case of clothing and footwear (43%), jewelry (41%), and catering services (38%); the lowest level was in the case of household appliances (16%).

Tailored customization was expected in furniture/home and garden equipment (37%), motorization/passenger cars (32%) and electronics (31%), and the least was in toys (25%).

The customized standardization respondents expected to buy electronics (36%), toys (32%), and beauty/cosmetics (31%); in the case of clothing and shoes, this was only 16%.

Pure standardization respondents were interested in their relationships to products of general use (35%), beauty/cosmetics (28%), and household appliances (27%). The least frequently mentioned answer in the case of pure standardization was catering services (10%).

According to customers in the case of pure customization, the price can be up to 50% higher; for tailored customization, up to 25% higher; and for customized standardization and pure standardization, up to 10% higher.

In most cases, customers accept the need to wait longer for a personalized product. When buying a car, they are willing to wait for up to 6 months for their order to be processed. They expect orders of up to a week for clothing and shoes, jewelry, beauty/cosmetics, toys, and products of general use. Customized catering services should be provided for up to a few hours.

The survey also examined the respondents' opinions on the preferred ways of contact between the consumer and the producer when creating a personalized product. The preferred forms of contact are direct personal contact with a manufacturer's representatives (52%), use of e-commerce channels (B2C, active website) (46.51%), and e-mail (42.25%). None of the answers proposed in the survey received a significant number of responses with the indication "definitely no."

The dynamic development of ICT networks, changing lifestyles, and forms of work leading to the networking of the society are the primary reasons for the growing number of network users. The vast majority of the respondents declared that they use the Internet every day (90%); only 1.2% of the respondents use it once a week, and about 5.5% declared that they use the Internet only a few times a week.

Most often, the Internet is used for maintaining relationships with other social media users (81%), receiving and sending e-mails (60%), making purchases of goods in online stores (58%), using Internet banking (54%), searching for news (52%), consuming video, movies, online TV, or online radio (51%), participating in education, training, or online courses (43%), Internet communicators (37%), browsing websites for entertainment or pleasure (34%), and organizing and collecting information about travel (23%) (Figure 11.1). The development of mobile technologies affects the dynamic growth of Internet users and allows for the development of various Internet services. The Internet is becoming one of the key media channels that shapes public opinion and consumer attitudes.

These studies are currently being conducted in the Czech Republic, Greece, Hungary, and Slovakia. As the first results show, the respondents' preferences are very similar to those presented in this study.

Findings and Recommendations

A clear direction of the changes in building the competitiveness of modern companies is the creation of value by customers. This means that customers take actions that affect the final composition of value that they themselves or other clients receive from a company—this can be considered to be the trend for Marketing 4.0.

As confirmed by the results of the conducted research, we are currently dealing with customer-centricity. In the face of dynamic market changes, a business cannot be indifferent to the growing expectations of

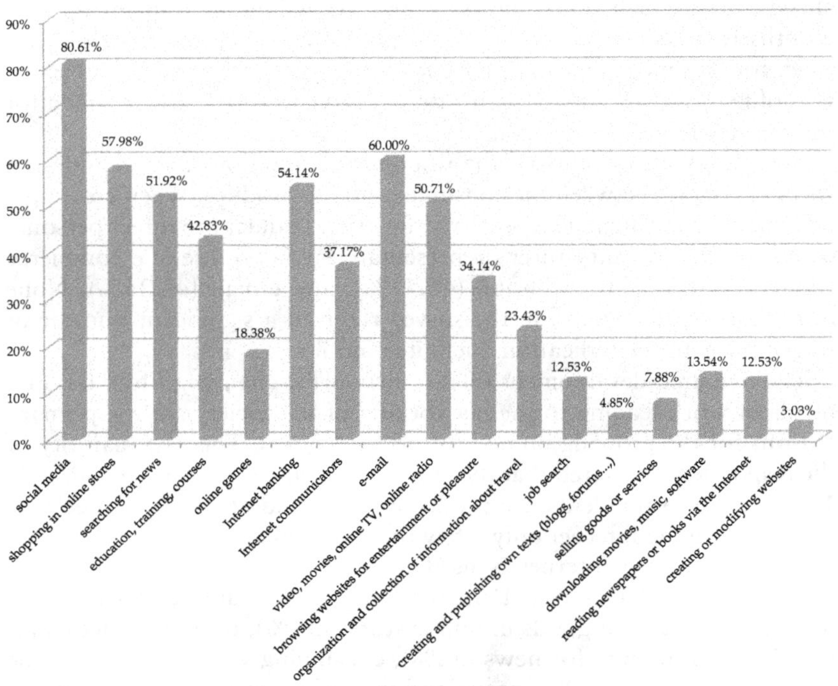

Figure 11.1 Key reasons for using Internet by modern consumer

Source*:* authors' elaboration

consumers. These expectations not only apply to products but also apply to prices, promotions, added values, or building relationships with an enterprise.

Consumers expect a personalized product, as such a product makes them feel special (according to the respondents). Most often, respondents indicate their willingness to buy personalized clothing and shoes, electronic devices, catering services, personalized accessories, and jewelry. A high level of expectations in terms of personalization is characterized by home and garden furnishings and cosmetics. Consumers are only willing to pay more for personalizing a product in exceptional cases; this means that developing the idea of Industry 4.0 to enable a high level of customization at a low level of production costs is necessary.

Consumers are interested in buying personalized products due to their uniqueness, such products make the consumer feel special, and they are convinced that such products are of higher quality. They often indicate that such products make excellent gifts. They value their influence on product features. On the basis of the conducted research, consumers are characterized by different levels of expectations regarding their involvement in the process of creating a personalized product. For clothing, shoes, jewelry, and catering services, they expect pure customization—they want to influence the design of new products. Home and garden equipment, motorization (cars), and electronics have a slightly lower level of consumer expectations for a personalized product. Pure standardization is expected in everyday-use products. For all product groups, consumers accept customized standardization that corresponds to current market trends. Consumers accept higher price levels with higher levels of customization (pure customization and tailored customization). In the case of customized standardization, they accept a small price increase. In the case of pure standardization, they do not accept a higher price for a product.

By analyzing the behavior of modern customers, Marketing 4.0 will be a mix of online and offline consumer experiences appearing along the entire path that leads to purchases. Due to the available data, predictive models, and extensive personalization options, it is possible to reach the right customer with the right offer at the right time—not counted in days but in seconds. Consumers indicated different acceptable waiting times for personalized products. The shortest declared waiting time applies to catering services and products of general use. For more complex and more expensive products, consumers accept longer waiting times (cars, electronics, clothing/shoes, and jewelry).

Digital technology brings a lot of change to concepts, instruments, and marketing. So, in which direction will the marketing activity evolve today? Will the available data and technologies such as the Internet of Things and artificial intelligence completely change marketing? Is there still room for creativity in Marketing 4.0?

The considerations presented in this chapter show the need to build relationships and cooperation between the consumer and the producer in the process of designing, manufacturing, and evaluating products—especially those that are personalized. This requires further in-depth research into the development of effective tools, platforms for the exchange of information between the consumer and producer, analysis of the data obtained from consumer activity on the Internet, and the use of artificial intelligence in virtual customer service.

This chapter is addressed to the people involved in marketing in various industries. The results of the research show the expectations of customers in the Fourth Industrial Revolution, which will help facilitate the design of products and the creation of all marketing activities.

References

Azmi, A. N., Kamin, Y., Noordin, M. K., & Nasir, A. N. M. (2018). Towards industrial revolution 4.0: Employers' expectations on fresh engineering graduates. *International Journal of Engineering & Technology*, 7(4/28), 267–272.

Barry, D., Helstrom, B., & Potter, J. (2011). Information management and related technology. In *Asset management excellence: Optimizing equipment life-cycle decisions* (p. 89). Taylor & Francis Group.

Berger, R. (2016). The Industrie 4.0 transition quantified. How the fourth industrial revolution is reshuffling the economic, social and industrial model. Roland Berger.

Choo, C., Detlor, B., & Turnbull, D. (2013). Web work: Information seeking and knowledge work on the world wide web. *Springer Science & Business Media*, *1*.

Chui, M., Löffler, M., & Roberts, R. (2010). The internet of things. *McKinsey*, *2*(47), 1–9.

Donnelly, C., Simmons, G., Armstrong, G., & Fearne, A. (2015). Digital loyalty card 'big data and small business marketing: Formal versus informal or complementary? *International Small Business Journal*, *33*(4), 422–442.

Dunn, K. (2015). Globalization and consumer: What marketer needs to know. *The Neumann Business Review*, 16–30.

Fogliattoa, F. S., da Silveirab, G., & Borensteinc, D. (2012). The mass customization decade: An updated review of the literature. *International Journal of Production Economics*, *138*(1), 14–25.

Gordon, J. (2013). *Big data, analytics and the future of marketing and sales*. Forbes. www.forbes.com/sites/mckinsey/2013/07/22/big-data-analytics-and-the-future-of-marketing-sales/#1a9591ae344d

Grabowska, S., Gajdzik, B., & Saniuk, S. (2020). The role and impact of industry 4.0 on business models. In *Sustainable logistics and production in industry 4.0. EcoProduction* (pp. 31–50). Springer.

Imran, M., Hameed, W., & Haque, A. (2018). Influence of industry 4.0 on the production and service sectors in Pakistan: Evidence from textile and logistics industries. *Social Sciences*, *7*, 246.

Jara, J., Concepción Parra, M., & Skarmeta, A. (2012). *Marketing 4.0: A new value added to the marketing through the internet of things*. 2012 Sixth International Conference on Innovative Mobile and Internet Services in Ubiquitous Computing, Palermo, pp. 852–857.

Jiménez-Zarco, A., Rospigliosi, A., Pilar Martínez-Ruiz, M., & Izquierdo-Yusta, A. (2017). *Marketing 4.0: Enhancing consumer-brand engagement through big data analysis*. IGI Global—All Rights Reserved.

Korena, Y., Shpitalnib, M., Guc, P., & Hu, S. (2015). Product design for mass-individualization. *Procedia CIRP*, Elsevier, *36*, 64–71.

Kotler, P., Kartajaya, H., & Setiawan, I. (2016). *Marketing 4.0: Moving from traditional to digital*. Hardcover.

Lies, J. (2017). Marketing 4.0 als Mittelstandskommunikation. In *Die Digitalisierung der Kommunikation im Mittelstand. essentials*. Springer.

Longo, F., Nicoletti, L., & Padovano, A. (2017). Smart operators in industry 4.0: A human-centered approach to enhance operators' capabilities and competencies within the new smart factory context. *Computers & Industrial Engineering*, *113*, 144–159.

Maresova, P., Soukal, I., Svobodova, L., Hedvicakova, M., Javanmardi, E., Selamat, A., & Krejcar, O. (2018). Consequences of industry 4.0 in business and economics. *Economies*, *6*, 46.

Müller, J., Kiel, D., & Voigt, K.-I. (2018). What drives the implementation of industry 4.0? The role of opportunities and challenges in the context of sustainability. *Sustainability*, *10*, 247.

Piccarozzi, M., Aquilani, B., & Gatti, C. (2018). Industry 4.0 in management studies: A systematic literature review. *Sustainability*, *10*, 3821.

Ranjan, K., & Read, S. (2016). Value co-creation: Concept and measurement. *Journal of the Academy of Marketing Science*, *44*(3), 290–315.

Rospigliosi, A., & Greener, S. (2015). *Leading issues in social media research*. Academic Publishing Ltd.

Sashi, C. (2012). Customer engagement, buyer-seller relationships, and social media. *Management Decision*, *50*(2), 253–272.

Suomala, P., Sievänen, M., & Paranko, J. (2002). Customization of capital goods—implications for after sales. In *Moving into mass customization*. Springer.

Want, R., Schilit, B., & Jenson, S. (2015). Enabling the internet of things. *IEEE Computer*, *48*, 1.

Wortmann, F., & Flüchter, K. (2015). Internet of things. *Business & Information Systems Engineering*, *57*(3), 221–224.

Yang, S., Kaminski, J., & Pepin, H. (2018). Opportunities for industry 4.0 to support remanufacturing. *Applied Sciences*, *8*, 1177.

12 Needs of Competency in Industrial Enterprises in Industry 4.0 Development Perspective

Waldemar Jędrzejczyk

Introduction

The technological environment of enterprises is very dynamic. In the last decade, this was mainly characterized by an increase in the volume of available data, the growth of computational capabilities, the wide use of mobile communications for the transmission of data from devices, progressing automation, the popularization of robots, and the development of digital channels of access to consumers. These trends are important enough in regard to production to be referred to as the Fourth Industrial Revolution concerning production, based on which the Industry 4.0 concept was developed. This leads to numerous changes in enterprises, which mainly refer to manufacturing processes. The most important trends include the following (Porter & Heppelmann, 2014; Kagermann, 2014, p. 607; PricewaterhouseCoopers, 2016, p. 4; Zawadzki et al., 2016; Bendkowski, 2017, pp. 21–24; Siemens, 2017, pp. 13, 134; Jadertrierveiler et al., 2019; Vrchota et al., 2020, p. 1):

- Development of new business models, services, and products;
- Mass personalization;
- Change of model of production and customer relationships;
- Change of value creation chains;
- Changes in the work area (demand for workforce, role of man in an intelligent factory, work organization, and environment), which will require a change of competencies.

The Industry 4.0 concept will radically change the model of functioning of all organizations (industrial enterprises in particular) (Propris & Bailey, 2020). This transformation will have a holistic nature. Organizational structures, manners of operation, and the work environment will change in a most significant way. Radical change will also concern the human role in the organization (ManpowerGroup, 2018; Paris & Giustiniano, 2018, pp. 256–257; Di Nardo et al., 2020). New positions and

DOI: 10.4324/9781003186373-12

professions will appear. The scope of responsibility will change. New competencies will be desirable (Raskino & Waller, 2015).

The issue of employee competencies in the context of the Industry 4.0 concept will be the principal subject of the matter of concern in this chapter, and the main purpose will be to present the basic trends of changes in the production work area caused by technological progress as well as to define the most desirable competencies in industrial enterprises in the perspective of the next few years. Such a research issue is characterized by a substantial level of difficulty. This difficulty results from both a lack of possibilities of performing a conclusive assessment of the influence of the realization of the Industry 4.0 concept on changes in the work area and the complexity of the competence issue. Currently, competencies—broadly understood—mean all attributes of an employee, the possession and skillful use of which contribute to the effective and efficient execution of company goals (Jędrzejczyk, 2013, p. 88).

Issues related to foreseeing the needs for competence in future enterprises (where business models will be shaped by new technologies) are being handled by numerous scientific and research centers worldwide. This chapter refers to the results of tests conducted by international consulting and advisory companies (Deloitte, ManpowerGroup, Boston Consulting Group [BCG], Allianz Global Corporate & Speciality [AGCS], and PricewaterhouseCoopers [PwC]) as well as global suppliers of modern technological solutions (Siemens, DELL Technologies, and the Institute for the Future).

Change of Trends in Production Work Area Triggered by Implementation of Industry 4.0 Concept

At present, the influence of the implementation of the Industry 4.0 concept on changes in the production work area cannot be unequivocally assessed. Elaborations concerning the Industry 4.0 concept allow us to distinguish two opposite streams of changes referring to employees and jobseekers: positive and negative. The foreseen stream of positive changes foresees that reindustrialization processes will influence the creation of a new better work environment aimed at the human factor and human needs (Institute for the Future, 2017). People will still be the most important asset of an organization. They will still supervise the production processes and machines, with the support of intelligent systems (Windelband, 2014, pp. 155–156). An increase in employment is also foreseen (BCG, 2015). Whereas the expected stream of negative changes is mainly foreseen in the progressing phenomenon of replacing people in production processes with new technological solutions, both in physical work and mental effort. New technological solutions are based on statistics, algorithms, the theory of probability, big data analyses, machine learning, and artificial intelligence. The scenario of comprehensive digitalization and automation foresees that the responsibility for steering the processes in intelligent enterprises will be taken over by socio-technical production systems. Highly qualified employees will be needed for system installation and maintenance, and most employees will be delegated to performing only simple tasks. Sceptics foresee a dehumanization of enterprises that are connected with staff reductions, decreases in salaries, and (simultaneously) increases in competency requirements (Schlung et al., 2014).

Both change streams assume that mass automation and robotization processes will first cover simple and routine activities. This will result in significant growth of demand for a highly qualified workforce, with a simultaneous decrease of demand for less qualified employees. The increase of competency requirements will not only result from implementing new technologies (specialist knowledge required) but will also be the effect of expanding the scope of responsibilities on existing job positions—combining technical and management activities in the production area. The decrease of simple tasks in the production process will be somewhat compensated by the establishment of new jobs in the planning and maintenance areas (Kurz, 2014; Bonekamp & Sure, 2015). However, it is expected that a significant number of less qualified employees who are not able to adjust their competencies to new challenges will lose their jobs (Spath et al., 2013).

The lack of possibilities for performing a conclusive assessment of the influence of the Industry 4.0 concept on changes in the workforce area is also confirmed by the latest research. Table 12.1 presents the results of research conducted by Deloitte Insights in regard to the probability of the occurrence of defined scenarios of changes in the work area caused by the development of Industry 4.0.

The opinions of senior management are split; e.g., with regard to the formation of relationships between organizations and employees, most respondents agree that these relationships will become weaker, but a significant number of respondents claim otherwise.

Need for Competence in Future Enterprises

The progressing digital transformation of industrial processes combined with their comprehensive automation and robotization will be a trigger for substantial changes in terms of the desired competencies, not only professional and technical competencies but also social and conceptual (Botthof & Hartmann, 2015). This is obvious for everybody; however, such unanimity does not exist in assessing which particular competencies will be needed. One of the reasons for this is that jobs and professions that do not exist yet will progressively be created.

Most forecasts regarding future competencies assume the need for developing competencies by the people employed in an industry

Table 12.1 The Fourth Industrial Revolution's impact on workforce (N=1,600 C level executives)

Which of the following statements is more true?		
No.	*Variant of response*	*Frequency of responses*
1	We are doing everything we can to create a workforce for the Fourth Industrial Revolution	86%
	We are not focused enough on creating a workforce for the Fourth Industrial Revolution	14%
2	We will need a complete rethinking of social/ labor contracts	56%
	The current labor laws and contracts will generally continue to work	44%
3	Our current education system will continue to work and prepare individuals for Industry 4.0	65%
	We will need a complete rethinking of the education system	35%
4	Our organization's relationships with our workforce will trend toward contractual, temporary, and/or ad hoc employees	61%
	Our organization's relationships with our workforce will trend more toward sustained full-time employment	39%
5	The skills that we require of our employees will evolve much more rapidly than how they do today	56%
	The skills that we require of our employees will evolve in a relatively similar fashion to those of today	44%
6	The majority of our workforce can be trained to have the skills we require	61%
	We will need to hire new/different people to possess the skills we require	39%
7	We will face social upheavals/increased income inequality	13%
	The Fourth Industrial Revolution will lead to more equality and stability	87%

Source*:* Deloitte (2017)

(Chenoy et al., 2019; Szafrański, 2019, p. 1007). It is due to the substantial growth of the complexity of the production processes (Kagermann et al., 2013):

- the change of the paradigm from "centralized production" to "decentralized production";
- the creation of new business models in terms of participation in online production processes, where employees will be able to flexibly configure the time of their physical presence at the workplace due to the possibility of remote communication with devices and steering their operation.

The research and analysis conducted by various R&D entities demonstrate that enterprises will mostly seek employees with very high competencies in the areas of specialist knowledge who may be defined as Industry 4.0 specialists—most of whom predestined to be engineers (Jin & Choi, 2019). They will have a particularly important function at the stage of implementing the Industry 4.0 concept—designing and implementing a new business model. The most important competency that they should have is interdisciplinarity, meaning the ability to combine and use knowledge from various technical areas (mostly automation, robotics, mechatronics, and Information Technology) as well as various management areas (mostly managing production processes, human resources management, and communications) (Hecklau et al., 2016). An Industry 4.0 specialist should therefore typically have both professional and engineering competencies as well as management competencies—cross-interpersonal competencies that are grouped into technical, personal, and social competencies from the classical point of view (Table 12.2), whereas technical and personal competencies are the leading competencies.

There are also other approaches to grouping competencies that should characterize an Industry 4.0. specialist. For example, Deloitte Insights distinguished five key categories of competencies, such as the following (Deloitte, 2018a, p. 8; *N*=400 companies):

1. Technology and computer skills;
2. Digital skills;
3. Programming skills for robots and automation;
4. Working with tools and techniques;
5. Critical thinking skills.

These will condition the effectiveness of implementing the Industry 4.0 concept to the greatest extent in industrial enterprises.

What is significant is that, in all of the analyzed cases, a relatively high importance was also assigned to general competencies and the so-called soft competencies; without these, it is impossible to work effectively in technical professions (Jędrzejczyk, 2019, p. 21; Siemens, 2019, p. 25).

The foreseen competency needs in future enterprises may be put in a holistic order (e.g., in the form of a model). The key competencies presented by the "Engineering Competency Model" should characterize an Industry 4.0 specialist. This model has six competency levels (CareerOneStop, 2017):

Level 1. Personal effectiveness competencies.
Level 2. Academic competencies.
Level 3. Workplace competencies.
Level 4. Industry-wide technical competencies.

Table 12.2 Suitability of competencies in perspective of requirements of modern innovative industry (*N*=200 companies)

Categories of competencies	*Frequency of responses (Sum of positive indications)*
Technical skills, which require the knowledge and understanding of the production process, including the ability to handle and configure production steering systems and the ability to intervene in case of an emergency, among others	93%
Interpersonal skills: analytical thinking, problem-solving, creative approach, openness to innovation, liability, handling the decision-making process, good time organization, openness to constant learning and knowledge sharing	89%
Social skills connected with communications and cooperation with others (including representatives of other cultures), understanding needs, leadership, initiating and maintaining business contacts	81%
Skills related to data, which require knowledge about data collection, storage, protection, and analysis, skills to make data-based decisions as well as skill in terms of programming, building algorithms, application of digital tools	66%

Source: Siemens (2019)

Level 5. Industry-sector functional areas.
Level 6. Job-specific competencies.

The pyramid of competencies was prepared by The Employment and Training Administration (ETA) in cooperation with the American Association of Engineering Societies (AAES). In this model, the basic categories of competencies that are required in all organizations include personal competencies, basic technical & IT competencies, and social competencies—Levels 1–3. Level 4 includes other competencies required in all industrial enterprises—broadly understood technical competencies. The last, highest levels of the pyramid include key competencies for sector specialists.

This is a model characterized by a large degree of generality—it demonstrates the basic types/categories of necessary competencies. Similar model solutions were formulated in reference to previous economic trends related to the development of technology and the increasing importance of intangible resources. The proposed model of engineering competencies coincides with the recommendation of the European Parliament and of the Council of the European Union on key competencies, which should be developed in the lifelong learning process. This recommendation covers eight key competencies that are essential for operating in a function-based economy. These include the following competencies (EU Council, 2018, pp. 7–8):

1. Literacy competence;
2. Multilingual competence;
3. Mathematical competence and competence in science, technology, and engineering;
4. Digital competence;
5. Personal, social, and learning to learn competence;
6. Citizenship competence;
7. Entrepreneurship competence;
8. Cultural awareness and expression competence.

The selected competencies include technical, personal, and social competencies.

Numerous elementary competencies may be defined within the scope of the individual key categories of competencies. Elementary competencies (which are very likely to be required in the Industry 4.0 work environment—most frequently indicated by various sectoral researchers) may include the following (Kurz, 2014, pp. 107–108; Hecklau et al., 2016; Sallati et al., 2019, p. 207; Siemens, 2019, pp. 9, 66):

- Ability to solve complex problems;
- Ability to learn at workplace;
- Ability to act flexibly;
- Ability to cooperate;
- Ability of analytical and logical thinking;
- Skill to cope in unforeseen situations;
- Creativity—in areas of managing communications, information, digitalization;
- Ability of self-organization and self-navigation;
- Competencies in the area of data protection and cyber-safety;
- Ability to assess the usability of new technologies;
- Ability to implement new technologies.

Changes in competency requirements will concern the entire enterprise personnel to a smaller or larger degree. It is foreseen that the development and implementation of the Industry 4.0 concept will require the most significant change of the competence profile of engineering personnel—the profession of engineer will evolve toward fulfilling the role of a change leader.

Contemporary Competence Problems in Organizations

One of the main factors that hampers the pace of innovation and the implementation of the Industry 4.0 concept includes staff-related limitations (Allianz Global Corporate & Speciality, 2019), which result from the insufficient knowledge of enterprise management concerning any ongoing changes and a shortage of Industry 4.0 specialists (Siemens, 2017, p. 26, 2019, p. 9). Entrepreneurs emphasize the need to engage new qualified specialists—mostly engineering personnel.

Enterprises are already facing the problem of finding employees with the required competencies, not only in the area of these challenges resulting from the development of the Industry 4.0 concept. More and more frequently, there is a lack of candidates who are willing to fill a vacancy. If there are candidates, then they are usually lacking adequate professional experience or have insufficient hard competencies or personal competencies. One of the barriers in engaging the appropriate people is related to a candidate's very high financial expectations (ManpowerGroup, 2018; *N*=39,195).

This problem will continue to grow. The competency gap in the area of the challenges related to Industry 4.0 (the differences between the number of job offers and the number of properly qualified personnel) will escalate (Siemens, 2017, p. 27). It has been foreseen that, during the coming decade, production enterprises will have a problem with finding approximately 2.4 million employees, which will constitute approximately 50% of their job offers (Deloitte, 2018b, p. 3). It will be most difficult to find employees with competencies in the area of digital technologies and production processes as well as operational managers (Deloitte, 2018b, p. 4; Siemens, 2019, p. 11).

The factor that will certainly make it easier to fulfill the requirements of the new job market will be Generation Y's entry onto the market (also referred to as Millennials). This is a generation of educated people with ease and fluency in modern IT technologies who are connected to the global network (usually via laptops, tablets, and smartphones).

Discussion

The Industry 4.0 concept is relatively new. It is innovative; the current stage of Industry 4.0 implementation may be seen as the pilot phase. Therefore, it seems justified to treat it as part of the category of a start-up undertaking and refer to the start-up life cycle. The first and most important stage in a start-up life cycle is conceptualization. This is a period of seeking and defining a problem as well as defining the solution (Picken, 2017, p. 588). Therefore, most needed at the moment are visionaries and change leaders who will design changes in enterprises and create conceptual models. They must possess knowledge, both in regard to the possibilities of new technological solutions as well as with the entire group of value-creation chains. This requires skills to undermine stereotypes and apply unconventional solutions. New solutions are usually placed on the opposite pole in regard to an existing solution (e.g., linear dependencies vs. network dependencies, centralization of process/management vs. decentralization of process/management, fixed working hours vs. flexible working hours).

The Industry 4.0 concept requires skills to combine hard and soft competencies; this means that most desired are specialists with very high competencies in both categories. Here, a difficulty arises that is related to human nature. People may generally be divided into two groups: "objectivists" and "humanists." An objectivist has a better-developed left cerebral hemisphere and is naturally predisposed to learning technical subjects, and a humanist has a better-developed right cerebral hemisphere and natural predispositions to learning humanist subjects. There are not many people with both predispositions at high levels. The natural competency gap is eliminated through team activities; therefore, competencies such as the ability to cooperate with a team and communication skills are of such great importance.

The main categories of competencies that should characterize employees of an Industry 4.0 enterprise have actually not changed in terms of those that used to be expected and are currently expected. Generally, competencies are divided into professional competencies and personal competencies. Professional competencies are defined in regard to a given job position. They are directly connected with the execution of tasks (e.g., technical skills, knowledge of the industry). These are defined as hard competencies. Personal competencies are related to the personal attributes of employees that are contributed to their professional roles; they concern behavior. Personal competencies are referred to as soft competencies.

Hard competencies are usually identified with technical competencies and soft competencies—with conceptual and social competencies. However, this manner of grouping competencies is not new. Even at the beginning of the 20th century, H. Fayol distinguished three basic analogical types of skills (technical, social, and conceptual) that are required for effective action (Fayol, 1949); this concept was developed and promoted by R.L. Katz (Katz, 1955, pp. 33–42). A natural question arises on whether this concept has evolved in any aspect. The answer is simple: yes. H. Fayol's concept concerned the requirements placed exclusively before management. In the Industry 4.0 development perspective, the requirement of having distinguished competencies concerns not only management but also the engineering staff—Industry 4.0 specialists.

The transformation of organizations determined by technological progress known as the Industry 4.0 concept today is inevitable. In order for it to be a factor of success rather than of failure, the following are necessary:

1. Not to be afraid of introducing changes and new technologies—an enterprise's internal processes should be digitized, automated, and robotized as much as possible;
2. When creating a new work environment, define the activities that will be performed by people and the activities that will be performed by machines in order to properly allocate the roles and tasks;
3. Shape the work environment in such a way that the domain of the Industry 4.0 concept is the increasingly better integration between man and technology rather than competing with technology for jobs;
4. Ensure the adequate portfolio and competence potential of an enterprise so that technical engineering competencies (competencies in the field of expertise: automation, robotics, mechatronics, IT and telecommunications) and managerial interdisciplinary competencies (most desirable: entrepreneurship, creativity, communicativeness, ability to work in a group) are skillfully integrated;
5. Focus equally on the current and prospective competence needs of an enterprise—try to identify the competencies of the "future" and develop them at the different stages of professional development.

Shaping competencies that are determined by the implementation of the Industry 4.0 concept requires systemic actions. In this process, an important role is played not only by enterprises but also by the state, intermediary organizations, and research-development units. There is a desire for the greater integration of technical and economic universities; that is, the integration of general science competencies and personal competencies.

The considerations presented in this chapter can be extremely useful for all those who are interested and professionally engaged in the issues of organization and management—mainly, innovation, production engineering, and operational management (both in theory and organizational practice). They are useful particularly for entrepreneurs, managerial staff of production enterprises, decision-makers in production operations and HR departments, university authorities, both technical universities and social universities, as well as students who study technical and social subjects.

Conclusions

The key factor of success in implementing the Industry 4.0 concept in enterprises is the human factor—not technology *per se*, but people who can apply it effectively. Therefore, a current priority is given to activities that prepare both entrepreneurs and employees to fulfill the conditions of Industry 4.0. Entrepreneurs should be aware of the fact that digital transformation is a complicated and costly process that requires the engagement of many resources. However, success will be achieved only by those enterprises that can use technology in a creative manner. In turn, employees should realize that the key factor for remaining on the job market will be the ability to retrain and gain new skills in order to keep pace with any changes driven by digital transformation. At the moment, it can be foreseen that the most important competencies that an employee should have in a 4.0 economy include high content-related and specialist knowledge, an innovative approach, creativity in solving problems, the ability to share and promote knowledge, the stimulation of one's own development and organization, and multi-tasking.

The development of technology creates demand for new professions (for which the technologies become working tools). The demand for new competencies (finding employees with the desired competencies) as a result of technical and technological progress will be one of the most important issues in the HR area (which will have to be tackled by organizations in the future). The demand for key competencies in organizations is already huge, while these traits are difficult to find on the job market. Many organizations still apply HR practices that were used during the Third Industrial Revolution. There is a need for new management models to support the implementation of new business models and new work models.

The research results presented in this chapter concern holistic analyses; therefore, the resulting tendencies of changes are more important than the absolute values of the presented research results. It should be understood that, in terms of absolute values, there may be certain deviations for certain regions or countries (depending mainly on the level of their development and industrialization). For example, there are visible differences between the results of highly developing countries (Western European countries, the United States, Canada, Japan, Australia, New Zealand, and Israel) and the results of moderately developed countries undergoing an economic transformation (Central and Eastern European countries such as Albania, Bulgaria, the Czech Republic, Poland, Romania, Slovakia, Hungary, and the Baltic States). This elaboration skips cross analyses due to the limited volume of this chapter.

This elaboration constitutes a significant contribution to the further research and empirical analyses of this area since it is difficult to foresee how exactly the changes will proceed in the coming years. There are also no proven solutions in the management theory, which would indicate how to prepare an organization for the approaching changes. There is still no unequivocal response to the principal question regarding the Industry 4.0 development perspective: which competencies should characterize human resources to effectively take advantage of the technological change?

References

Allianz Global Corporate & Speciality. (2019). *Allianz risk barometer 2019.* www.agcs.allianz.com/assets/PDFs/Reports/Allianz_Risk_Barometer_2019.pdf

BCG—The Boston Consulting Group. (2015). *Industry 4.0: The future of productivity and growth in manufacturing industries.* https://image-src.bcg.com/Images/Industry_40_Future_of_Productivity_April_2015_tcm9-61694.pdf

Bendkowski, J. (2017). Zmiany w pracy produkcyjnej w perspektywie koncepcji "Przemysł 4.0". *Zeszyty Naukowe Politechniki Śląskiej, Seria: Organizacja i Zarządzanie, 112,* 21–33.

Bonekamp, L., & Sure, M. (2015). Consequences of industry 4.0 on human labour and work organisation. *Journal of Business and Media Psychology, 6*(1), 33–40.

Botthof, A., & Hartmann, E. A. (2015). *Zukunft der Arbeit in Industrie 4.0.* Springer.

CareerOneStop. (2017). *Engineering competency model.* www.careeronestop.org/competencymodel/competency-models/engineering.aspx

Chenoy, D., Ghosh, S. M., & Shukla, S. K. (2019). Skill development for accelerating the manufacturing sector: The role of 'new-age' skills for 'make in India'. *International Journal of Training Research, 17*(S1), 112–130.

De Propris, L., & Bailey, D. (Eds.). (2020). *Industry 4.0 and regional transformations.* Routledge.

Deloitte. (2017). *The fourth industrial revolution is here—are you ready?* https://www2.deloitte.com/content/dam/Deloitte/tr/Documents/manufacturing/Industry4-0_Are-you-ready_Report.pdf

Deloitte. (2018a). *2018 Deloitte and the manufacturing institute skills gap and future of work study.* www.themanufacturinginstitute.org/wp-content/uploads/2020/03/MI-Deloitte-skills-gap-Future-of-Workforce-study-2018.pdf

Deloitte. (2018b). *2018 Skills gap in manufacturing study.* www2.deloitte.com/us/en/pages/manufacturing/articles/future-of-manufacturing-skills-gap-study.html

Di Nardo, M., Forino, D., & Murino, T. (2020). The evolution of man-machine interaction: The role of human in industry 4.0 paradigm. *Production & Manufacturing Research, 8*(1), 20–34.

EU Council. (2018). *The council of the European Union recommendation of 22 May 2018 on key competences for lifelong learning.* https://eur-lex.europa.eu/legal-content/EN/TXT/PDF/?uri=CELEX:32018H0604(01) &from=LT

Fayol, H. (1949). General and industrial management. Pitman.

Hecklau, F., Galeitzke, M., Flachs, S., & Kohl, H. (2016). Holistic approach for human resource management in industry 4.0. *Procedia CIRP, ELSEVIER, 54,* 1–6.

Institute for the Future. DELL Technologies. (2017). *The next era of human. Machine partnerships. New report explores emerging technologies' impact on society & work in 2030.* www.iftf.org/humanmachinepartnerships

Jadertrierveiler, H., Sell, D., & Santo, N. D. (2019). The benefits and challenges of digital transformation in industry 4.0. *Global Journal of Management and Business Research: A Administration and Management, 19*(12).

Jędrzejczyk, W. (2013). *Intuicja jako kompetencja menedżerska w teorii i praktyce zarządzania przedsiębiorstwem*. Dom Organizatora.

Jędrzejczyk, W. (2019). Human-organization relation in the perspective of industry 4.0. In J. Trojanowska, O. Ciszak, J. M. Machado, & I. Pavlenko (Eds.), *Advances in manufacturing II. Vol. 1 — solutions for industry 4.0* (pp. 14–24). Springer.

Jin, S. H., & Choi, S. O. (2019). The effect of innovation capability on business performance: A focus on IT and business service companies. *Sustainability*, *11*(19), 5246.

Kagermann, H. (2014). Chancen von Industrie 4.0 nutzen. In T. Bauernhansl, M. ten Hompel, & B. Vogel-Heuser (Eds.), *Industrie 4.0 in Produktion, Automatisierung und Logistik*. Springer.

Kagermann, H., Helbig, J., Hellinger, A., & Wahlster, W. (2013). *Recommendations for implementing the strategic initiative INDUSTRIE 4.0: Securing the future of German manufacturing industry* (Final Report of the Industrie 4.0 Working Group). Forschungsunion.

Katz, R. L. (1955). Skills of an effective administrator. *Harvard Business Review*, *33*(1), 33–42.

Kurz, C. (2014). Industrie 4.0 verändert die Arbeitswelt. Gewerkschaftliche Gestaltungsimpulse für „bessere"Arbeit. In W. Schröter (Hrsg.), *Identität in der Virtualität. Einblicke in neue Arbeitswelten und "Industrie 4.0"*. Talheimer Verlag.

ManpowerGroup. (2018). *2018 Talent shortage survey*. www.manpowergroup.co.uk/the-word-on-work/2018-talent-shortage-survey/#report

Paris, A., & Giustiniano, L. (2018). Industry 4.0 and the emerging challenges to leadership. In F. Cantoni & G. Mangia (Eds.), *Human resource management and digitalization* (pp. 255–259). Routledge and Taylor & Francis Group.

Picken, J. C. (2017). From startup to scalable enterprise: Laying the foundation. *Business Horizons*, *ELSEVIER*, 60(5), 587–595.

Porter, M. E., & Heppelmann, J. E. (2014, November). How smart, connected products are transforming competition. *Harvard Business Review*.

PricewaterhouseCoopers. (2016). *Przemysł 4.0 czyli wyzwania współczesnej produkcji*. www.pwc.pl/pl/pdf/przemysl-4-0-raport.pdf

Raskino, M., & Waller, G. (2015). *Digital to the core. Remastering leadership for your industry, your enterprise, and yourself*. Routledge.

Sallati, C., de Andrade Bertazzi, J., & Schützer, K. (2019). Professional skills in the product development process: The contribution of learning environments to professional skills in the Industry 4.0 scenario. *Procedia CIRP*, *ELSEVIER*, *84*, 203–208.

Schlund, S., Hämmerle, M., & Strölin, T. (2014). *Industrie 4.0 eine Revolution der Arbeitsgestaltung—Wie Automatisierung und Digitalisierung unsere Produktion verändern wird*. Ingenics AG.

Siemens. (2017). *Od industry 4.0 do smart factory*. https://publikacje.siemens-info.com/webreader/00085-001733-od-industry-4 - 0-do-smart-factory-poradnik-menedzera-i-inzyniera/index.html#p=1

Siemens. (2019). *Smart industry polska 2019. Inżynierowie w dobie czwartej rewolucji przemysłowej*. Raport z badań. https://publikacje.siemens-info.com/pdf/594/Raport%20Smart%20 Industry%20Polska%202019.pdf

Spath, D., Ganschar, O., Gerlach, S., Hämmerle, M., Krause, T., & Schlund, S. (2013). *Produktionsarbeit der Zukunft—Industrie 4.0*. Frauenhofer Institut für Arbeitswirtschaft und Organisation.

Szafrański, M. (2019). *Threefold nature of competences in enterprise management: A qualitative model*. Proceedings of the 20th European Conference on Knowledge Management: Univerisdade Europeia de Lisboa, Lisbon, Portugal, September 5–6, 2019, Vol. 2, pp. 1006–1015.

Vrchota, J., Maříková, M., Řehoř, P., Rolínek, L., & Toušek, R. (2020). Human resources readiness for industry 4.0. *Journal of Open Innovation: Technology, Market, and Complexity*, *6*(3), 1–20.

Windelband, L. (2014). Zukunft der Facharbeit im Zeitalter "Industrie 4.0". *Journal of Technical Education*, *2*(2).

Zawadzki, P., & Żywicki, K. (2016). Smart product design and production control for effective mass customization in the industry 4.0 concept. *Management and Production Engineering Review*, *7*(3), 105–112.

13 Design Thinking for Industry 4.0 Career Design—How to Increase Professional Development Awareness for Future Enterprises' Human Resources

Jacek Jakieła, Joanna Świętoniowska and Joanna Wójcik

Introduction

According to one of many definitions, Industry 4.0 involves digitalizing manufacturing and business processes to create Smart Factories. The objectives are to increase productivity, improve efficiency, and develop new sales opportunities. It also offers new possibilities for career planning and development. People who are now at the first step of their higher educational ladder will meet a brand new job market after graduation in terms of job position requirements—the expected skills and knowledge areas.

The question is how to start planning one's professional development as early as possible and start learning as well as developing valuable skills while attending a university. This is not a trivial problem and maybe even considered to be a wicked problem. Usually, most young people simply do not know what makes them tick—what they really like to do—before they try it out and understand what it is about. Moreover, universities often have problems with fitting educational offers to real job market needs. The offer is rarely created in close cooperation with organizations that demand specific skills and knowledge. Therefore, the career planning and design approach have to take all of these aspects into consideration.

The aim of this chapter is to present the original approach developed by the authors in the BEAST (BE Aware STudent) project.[1] The project's outcomes can be used for career development during the educational process from its beginning. It enables young people to plan or replan the educational track at the university according to selected future job positions. It is also a change enabler for universities that can use the approach for getting the best fit between educational offers and job market requirements. The framework applied is the composition of the Design Thinking and business model thinking. The methodology is focused on top Industry

DOI: 10.4324/9781003186373-13

4.0 career paths (e.g., data scientist, business analyst, network engineer). However, the approach is very flexible and can be easily adapted to other career paths as needed as well as by HR departments of enterprises to develop the skills of staff members according to specific occupational requirements.

Career Planning and Most Important Competencies in Industry 4.0 Era

In the era of Industry 4.0, the approach to career planning is changing rapidly. An employee who wants to succeed in the labor market should become a one-person business enterprise (Weick, 1996, p. 54). This requires the fast and aware development of skills and knowledge. The path to success goes through continuous learning and improving qualifications, very often in the non-formal education system. According to Bohdziewicz (2010, p. 46), the career development model includes the self-management of one's career according to the rules imposed by the labor market, the autonomous building of employability and market attractiveness, and developing a professional identity. Prospective employees should define themselves not in terms of lifelong positions in specific companies but rather in their professional identities. This is caused by the changing form of employment and growing career mobility (Kornblum et al., 2018, p. 657). The full-time employment model has been replaced by part-time, temporary, and on-call models (Kergroach, 2017, p. 7).

An Industry 4.0 worker (Gracel & Stoch, 2016, p. 125) should no longer focus only on his/her narrow specialization but also should be able to understand concepts and ideas from other fields. It is possible to identify common key competencies and domain-specific ones (Simic & Nedelko, 2019, p. 1, 295). Deep technical knowledge is still a valuable asset, but new areas have appeared (see Table 13.1) (Eberhard et al., 2017, p. 53).

Important competencies can also be divided into such categories as technical, managerial, and social (Łupicka & Grzybowska, 2018, p. 253) and named in various ways (e.g., efficiency orientation, research skills, lifelong learning, flexibility, etc.). A full list of the names used by researchers can be found in Dos Santos and Benneworth (2019, p. 309).

Discovering Professional Identity With Design Thinking—Designing Professional Way Forward

Career Planning Is Wicked Problem

The career-planning process is overwhelming for most people; however, this is especially true for students who have little professional experience and do not have a vision for their lives after graduation. Problems are caused by a low awareness about shaping one's career path and the

Table 13.1 Skill portfolio for university training

Abilities	*Basic skills*	*Cross-functional Skills*	
Cognitive skills	**Content skills**	**Social/ Interpersonal skills**	**Resource management skills**
Cognitive flexibility Creativity Logical reasoning Complex problem-solving Mathematical reasoning Visualization Troubleshooting Analytical Skills (statistics) **Personal/mental abilities** Knowledge in psychology Body language Resilience Intrapreneurial skills	Active learning Oral expression Reading comprehension Written expression ICT literacy **Process Skills** Active listening Critical thinking Monitoring self and others Interdisciplinary skills	Coordinating with others Emotional Intelligence Negotiation Persuasion Service oriented Training and teaching others Ethics and social responsibility Virtual collaboration Communication skills **System skills** Judgment and decision-making System analysis Change management and adaption Governance, risk management Compliance Entrepreneurial skills	Mgmt. of financial resources Mgmt. of material resources Mgmt. of financial resources People management Time management **Technical skills** Equipment maintenance and repair Equipment operation and control Programming Quality control Tech. and user-experience design New technologies (ICT, etc.) **Intercultural skills** Language skills Open mindset

Source: Eberhard et al., 2017, p. 54; after WEF, 2016

lack of ability to choose the areas on which to focus. The next factor is a weak correlation between study programs and labor market requirements. Solving these problems needs an innovative approach that will help students discover their passions and design their professional lives. The properly selected tools will allow one to recognize their personal strengths/weaknesses and draw their attention to skills for which they have talent. A reasonable approach is to try things out and learn through actions and from personal experience. Universities are constantly looking for tools to enhance a student's skills with the purpose of preparing them for their professional career. Design Thinking is one of the approaches that universities can adopt to help students discover their long-term goals for growth and career development.

Idea of Design Thinking

From the business perspective, Design Thinking is a human-centered problem-solving process (Melles et al., 2012, p. 162) that provides numerous possibilities for innovation development (Brown & Wyatt, 2010) and economic benefits. This concept merges different approaches: creative problem-solving, an out-of-the-box way of thinking, and a methodology that can be used in multidisciplinary settings (Tantiyaswasdikul, 2019, p. 47).

Several design thinking models have been adopted; they include from three to seven stages of the process. Tim Brown from IDEO defines the design thinking process as a system of overlapping spaces (Brown & Wyatt, 2010, p. 33): *inspiration* (finding the problem or opportunity that motivates the search for solutions), *ideation* (the process of generating and developing ideas), and *implementation* (testing of ideas in real-life scenarios and implementing the solution into people's lives). The Hasso-Plattner Institute of Design at Stanford (d.school) proposes the five following stages of the process (Henriksen et al., 2017, p. 142): *Empathise*, *Define*, *Ideate*, *Prototype*, and *Test*. The UK's Design Council proposed four distinct phases: *Discover*, *Define*, *Develop*, and *Deliver*, which is known as the *Double Diamond Model* (see Figure 13.1).

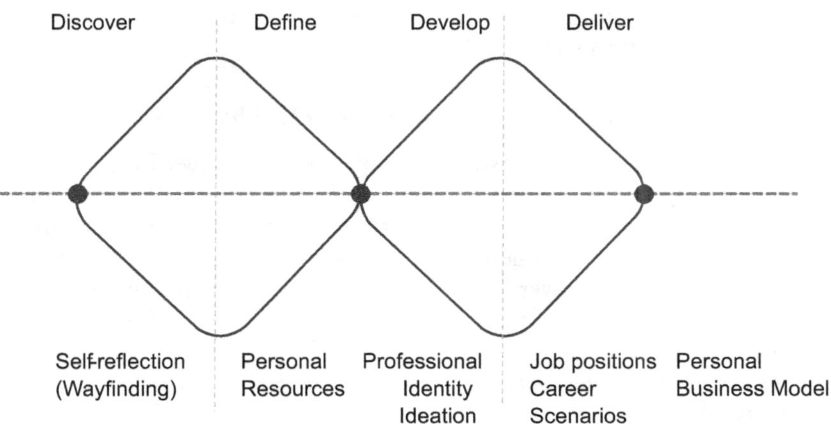

Figure 13.1 BEAST Double Diamond

Source: BEAST Project Intellectual Outcomes Documentation

The *Discover* stage helps us understand what the problem is. It involves interviewing and observing representatives of prospective user target groups. The aim of *Define* is to set up the challenge based on the knowledge elicited. The *Develop* stage is responsible for looking for inspiration as well as generating, reviewing, and selecting ideas. The *Delivery* phase consists of developing and testing different solutions on a small scale and rejecting those that do not work (Design Council, 2019).

Applying Design Thinking to Career Development

As a career planning is a wicked problem, universities can apply principles of the Design Thinking approach to support students in the process of generating and verifying new ideas in the fields of educational planning and career development. Design thinking can be considered to be an essential tool as well as an approach for problem-solving in the field of *professional identity awareness development*. The d.school at Stanford University was the first to incorporate the Design Thinking model into the professional development-planning process.

In the BEAST project, the application of Design Thinking concerns the whole process of iterative professional development, design, and redesign as needed. The main aim of such an application of the Design Thinking mindset is to provide students with a range of possible career paths from which they can select, design, test, and apply for professional development. Therefore, *the Double Diamond Model can serve as a reference model for an approach in which career-planning steps are based on the design thinking process, stages, and tools to enhance students' abilities to think about their professional futures and current educational contexts while attending university.*

First Stage—Self-Reflection/Wayfinding [DT *Discover*]

The first step is about understanding a problem to solve. In the case of the career-planning process, a student should perform a self-reflection at this stage. In the BEAST project, several techniques have been adopted (such as Holland's test to determine personality and work environment tendencies, lifeline discovery, good time journal, and find your WHY) to discover what constitutes the basement for well-being in the context of professional work. During the process of self-reflection, the student also identifies the main elements of his/her personal resources—personality, interests, abilities, skills, and knowledge areas.

Second Stage—Professional Life Principles/Personal Resources Definition [DT Define]

Next, the information collected during the previous phase is analyzed to better understand the professional life principles and personal resources that the student possesses; then, an assessment of their development level is done. The results of the first two stages should provide the answers to questions like "*What do I like to do?*", "*Do I like teamwork or rather individual work?*", "*What am I good at?*", "*What makes me happy?*", and "*In which activities I am in the flow state?*" These answers develop the awareness of the student on what he/she would like to do in his/her professional life.

Third Stage—Professional Identity Ideation [DT Develop]

In the *Ideation* phase, a range of solutions to the problem that has been defined is brainstormed. In this stage, a student can ideate several possible professional identities and job positions that are related to them. The process is supported by the IT Career Canvas Catalog (described in subsequent sections). Depending on the personal resources identified and their match with the resources needed for specific professions, the student can relatively easily select those that interest him/her.

Fourth Stage—Career Scenarios/Personal Business Model [DT Deliver]

In the frame of this stage, the student creates a few career scenarios and tests them in order to better understand his/her choices. Building a simple prototype is related to developing his/her personal business model canvas for the job position(s) selected. The testing can be based on several activities; the student can analyze the IT Career Canvas Catalog and find the gaps in his/her personal resources, listen to podcasts with professionals who hold specific job positions, finish one or two courses suggested in the catalog, or look at the requirements for any important certificates related to a profession. He/she can also interview some people working in that field to gain a deeper understanding of what the selected job position is about or attend a conference, seminar, or workshop. The testing phase should help the student gain insights and develop an awareness of any possible professional development directions.

The whole process is carried out in a flexible and non-linear iterative fashion. The prototypes are further refined based on the information collected in the test phase until a solution is reached. The student may explore a range of possible career scenarios before he/she finds the right one for him/her. *The benefit of the design process is that it provides a path to follow that allows for the exploration of new career options. This would help students to design a professional way forward.*

The crux of the BEAST approach is the reference model in the form of the IT Career Canvas Catalog and canvanizing process (which is presented in the following sections).

Canvanizing Job Market Requirements

Preparing future staff members for modern enterprises requires a comprehensive view of job market requirements. Such a view should include many aspects and characteristics of the ecosystem in which a prospective employee will be working. Any approach taken requires the proper technique to capture and map the job position business environment's most important elements. In the BEAST project, Business Model Canvas has been adopted (Osterwalder & Pigneur, 2010, pp. 18–46). It has been used to describe a work environment in the context of the key components that each enterprise has and uses when conducting business. Therefore, the business model concept is a driver of the understanding of employer requirements and career choices. According to Osterwalder and Pigneur, each business model can be expressed in terms of nine main components. These components can be grouped into three parts: the *customer-facing part*, *backstage*, and *financial sustainability part*.

The first part addresses the question of "*How does one build relationships with customer segments by creating and delivering value proposition?*" (segments: value proposition, channels, relationships, customer segments). The backstage or infrastructure part considers the *resources and activities used for creating and delivering the value proposition to the customers and partners that constitute a company's ecosystem*. The financial sustainability part regards the costs and revenues generated by an organization that is responsible for profit or loss (depending on the financial situation resulting from the invented business model). Understanding the business model of a company of which a specific employee will be a part enables one to identify the most valuable resources an individual must have to best fit the job market requirements.

In the approach developed in the BEAST project, we have made the *transition from a business model to a personal business model*. Each key human resource responsible for a specific element of an organization's value proposition possesses his/her own personal business model, which is an enabler for being an effective and efficient staff member. An employee's personal business model is isomorphic with the business model of the company he/she works for in terms of the model's structure and components' semantics. However, the context is different.

As in the case of Business Model Canvas, Personal Business Model also includes nine key building blocks (Clark & Osterwalder, 2012, pp. 53–74). However, their contents are not related to an organization in this case but to a staff member holding a specific job position (Figure 13.2). In this view, the *Customer Segments* component regards those who will pay to receive benefits or want to have the jobs done. They may be the stakeholders to which the staff member will report, internal customers or the organization's customers, business partners, and communities. This element together with the *Roles/Relationships* component

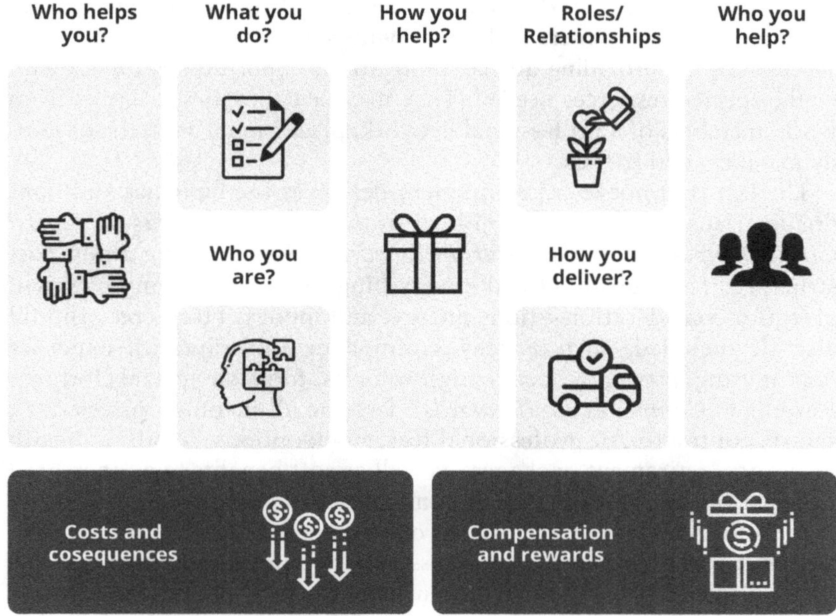

Figure 13.2 Personal Business Model Canvas

Source: Clark & Osterwalder, 2012, p. 55

enables us to understand the context of a staff member's interactions when working on a specific job position. The most important elements of the *Resources* component are interests, personality, abilities, and skills; it also includes knowledge, experience, and personal and professional contacts. Having identified all of these items for a specific job position, it becomes clear what are the real requirements for a prospective employee. Key activities describe the most important tasks a professional must carry out when getting jobs done. A detailed description of this element can provide a view of the typical activities done in a job position. If the activities are done effectively and efficiently, then the staff member can produce a unique value proposition that transforms into a company value proposition. The benefits provided by a staff member on a specific job position are delivered via the proper channels. Depending on workplace specificity, this can be done in many ways, such as submitting written reports, talking to people, uploading code to a development server, delivering oral presentations in person or online, and with vehicles (to physically deliver merchandise).

No job position is an island in an organization. Nowadays, the processes conducted in companies are organized around projects whose realizations are based on teamwork. Thus, it is good to know how a specific job position is related to the partners who can help. This is the responsibility of the *Key Partners* module in the Personal Business Model. Examples of key partners include all who support or help to do a job successfully by providing advice, motivation, opportunities for growth, or the specific resources needed. They may be colleagues, mentors from work, members of a professional network, professional advisers, or family members and friends.

The last (but not least) component describes the financial and non-financial sustainability of a job position with such areas as *Costs/Consequences* and *Compensation/Rewards*. When talking about cost structure, it is important to take everything a prospective employee will give into consideration—time, energy, and money. Hard costs should also be included, such as fees (training or subscription), expenses (commuting, travel, or socializing), vehicles, tools, or special clothing. Revenues (*Compensation/Rewards*) describe all income sources (e.g., salary, contractor, or professional fees, stock options, royalties, health insurance, retirement packages) as well as soft benefits (e.g., increased satisfaction, recognition, social contribution, flexibility).

In the BEAST project, the process of mapping job market requirements has been based on Personal Business Model Canvas. The final result is the IT Career Canvas Catalog that includes the personal business models of IT-related job positions that may be of interest to organizations that follow Industry 4.0 trends. The job positions considered as professions of the future in the IT industry (and selected in the BEAST project) are the following:

- Business analyst;
- Data scientist;
- Cybersecurity specialist;
- IoT specialist;
- Game designer;
- Network engineer;
- UX/UI designer;
- Enterprise architect/ERP specialist;
- Mobile developer/web developer.

An analysis of 1,460 announcements from the *LinkedIn* portal has shown that the phrases related to ICT that appear most frequently (excluding general knowledge) were machine learning, big data, and data analysis (Pejic-Bach et al., 2020, pp. 416–431). Therefore, the examples selected from the BEAST project are related to data science.

For each profession selected in the BEAST project, a detailed analysis has been conducted, and all key elements of the Personal Business Model Canvas were identified. The example of the reference Business Model Canvas for the data scientist profession is presented in Figure 13.3.

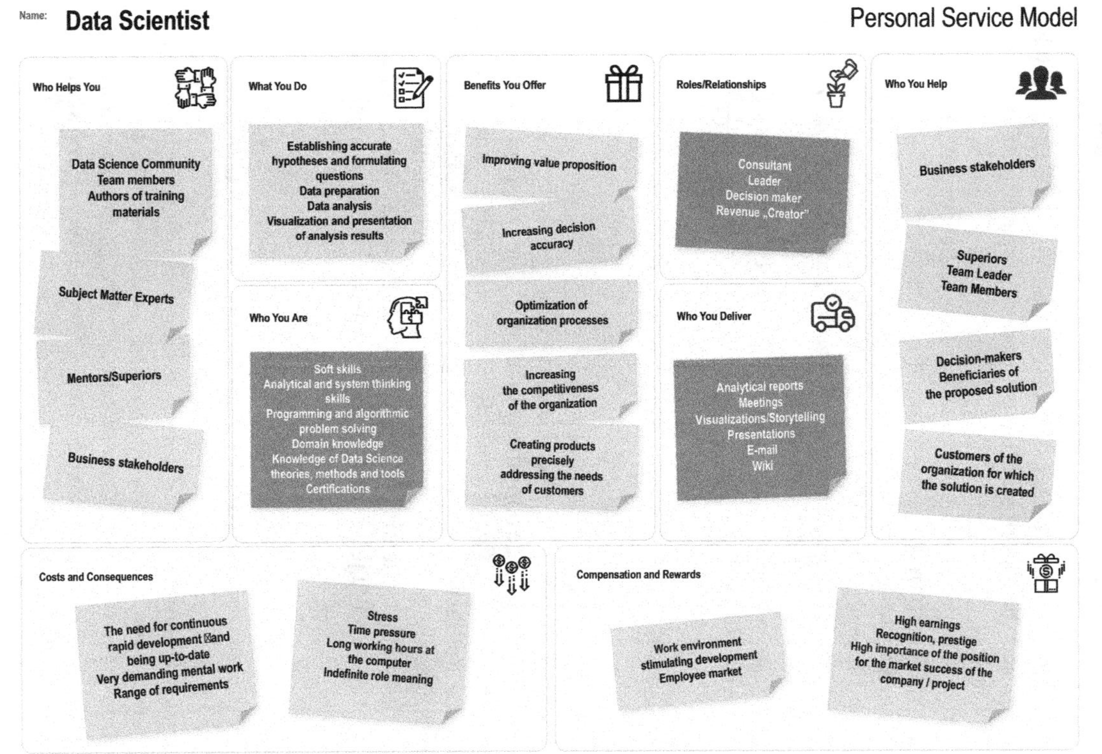

Figure 13.3 Reference Business Model for Data Scientist Job Position

Source: BEAST Project Intellectual Outcomes Documentation

Each canvas element has a detailed description of the job position requirements and the business context of the workplace.

The IT Career Canvas Catalog is regarded as the most important product of the project and may be used as the reference model for many career-planning activities. The three most important are as follows:

- *Finding Personal Resources Gaps*—what should I learn, and what skills should be developed to be better prepared for a specific profession?
- *Planning Career Development*—having specific personal resources for the job position that is best suited for me.
- *Redesigning the career path and career pivoting*—how to improve my workplace well-being by finding the best fit between my expectations, strengths, and current requirements of a job position.

The next sections of the chapter describe the first two areas of *IT Career Canvas Catalog* applications.

Planning Career Paths With Job Positions Reference Model

Identifying Personal Resources Gaps

As has been said, the *IT Career Canvas Catalog* may be used as a reference model for career development planning. The questions of "*What is one to learn?*" and "*What knowledge areas need to be developed better?*" may be addressed by finding the gaps in the *personal resources*. Assuming that someone knows the profession in which he/she would like to be a specialist, the first step is to analyze Personal Business Model Canvas for the selected job position. A hypothetical scenario is presented as follows:

> *Anna is a second-year student of computer science at the university. She likes numbers and data processing stuff. She has well-developed analytical and system-thinking skills and is sure that her future is in data science. However, she does not know which skills and knowledge areas should be further developed. She would like to focus on the university subjects that will provide her with opportunities to fill the personal resources gaps with the most important skills and knowledge. Of course, the importance should be related to the data scientist profession.*

In the first step, Anna can look at the *Personal Business Model Canvas* for *Data Scientist* job position and carefully analyze the *Who you are?* component that contains all of the skills, abilities, and knowledge areas that play an important role in this profession.

outcomes prepared for individual job positions) supports the process of exploring a range of possible career scenarios before finding the right one or enables one to find the best fit between occupation and the personal resources needed for a specific job position. However, the catalog requires testing in a wider group of people planning and replanning their professional careers. The next step planned is to develop a canvas that contains the key/universal competencies that an employee of Industry 4.0 should possess.

Note

1. The project was financed in the frame of the ERASMUS+ program (No. 2018–1PL01-KA203–051137) and led by the University of Information Technology and Management in Rzeszow in a partnership involving universities from Poland, Italy, and Portugal (2018–present).

References

Bohdziewicz, P. (2010). Modern career models: From bureaucracy to entrepreneurship (Współczesne kariery zawodowe od modelu biurokratycznego do przedsiębiorczego). *Zarządzanie Zasobami Ludzkimi*, *3–4*(74–75), 39–56.

Brown, T., & Wyatt, J. (2010). Design thinking for social innovation. *Development Outreach*, *12*(1), 29–43.

Clark, T., Osterwalder, A., & Pigneur, Y. (2012). *Business model you: A one-page method for reinventing your career*. John Wiley & Sons.

Design Council. (2019). www.designcouncil.org.uk

Dos Santos, E. F., & Benneworth, P. (2019). Makerspace for skills development in the industry 4.0 era. *Brazilian Journal of Operations & Production Management*, *16*(2), 303–315. http://dx.doi.org/10.14488/BJOPM.2019.v16.n2.a11

Eberhard, D. A., et al. (2017). Smart work: The transformation of the labour market due to the fourth industrial revolution (I4.0). *International Journal of Business and Economic Sciences Applied Research (IJBESAR)*, *10*(3), 47–66. ISSN 2408–0101, Eastern Macedonia and Thrace Institute of Technology, Kavala. http://dx.doi.org/10.25103/ijbesar.103.03

Gracel, J., & Stoch, M. (2016). Inżynierowie przemysłu 4.0: jak ich rozwijać? *Harvard Business Review Poland*. www.hbrp.pl/b/inzynierowie-przemyslu-40-jak-ich-rozwijac/1A0LUxCGY

Henriksen, D., Richardson, C., & Mehta, R. (2017). Design thinking: A creative approach to educational problems of practice. *Thinking skills and Creativity*, *26*, 140–153. https://doi.org/10.1016/j.tsc.2017.10.001

Kergroach, S. (2017). Industry 4.0: New challenges and opportunities for the labour market. *Foresight and STI Governance*, *11*(4), 6–8. http://dx.doi.org/10.17323/2500-2597.2017.4.6.8

Kornblum, A., Unger, D., & Grote, G. (2018). When do employees cross boundaries? Individual and contextual determinants of career mobility. *European Journal of Work and Organizational Psychology*, *27*(5), 657–668. doi:10.1080/1359432X.2018.1488686

Łupicka, A., & Grzybowska, K. (2018). Key managerial competencies for industry 4.0-practitioners', researchers' and students' opinions. *Logistics and Transport*, *39*.

Melles, G., Howard, Z., & Thompson-Whiteside, S. (2012). Teaching design thinking: Expanding horizons in design education. *Procedia: Social and Behavioral Sciences*, *31*, 162–166. https://doi.org/10.1016/j.sbspro.2011.12.035

Osterwalder, A., & Pigneur, Y. (2010). *Business model generation: A handbook for visionaries, game changers, and challengers*. John Wiley & Sons.

Pejic-Bach, M., Bertoncel, T., Meško, M., & Krstić, Ž. (2020). Text mining of industry 4.0 job advertisements. *International Journal of Information Management*, *50*, 416–431. https://doi.org/10.1016/j.ijinfomgt.2019.07.014

Simic, M., & Nedelko, Z. (2019). Development of competence model for industry 4.0: A theoretical approach. *Economic and Social Development: Book of Proceedings*, 1288–1298.

Tantiyaswasdikul, K. (2019). A framework for design thinking outside the design profession: An analysis of design thinking implementations. *Journal of Architectural/Planning Research and Studies (JARS)*, *16*(1), 45–68.

WEF. (2016). *The future of jobs: Employment, skills and workforce strategy for the fourth industrial revolution (executive summary)*. World Economic Forum. http://www3.weforum.org/docs/WEF_FOJ_Executive_Summary_Jobs.pdf

Weick, K. E. (1996). Enactment and the boundaryless career: Organizing as we work. *The Boundaryless Career: A New Employment Principle for a New Organizational Era*, 40–57.

14 Challenges and Problems in Managing Multigenerational Team in Era of Industry 4.0

Magdalena Maciaszczyk and Damian Kocot

Introduction

In the current economic reality, multigenerationality (i.e., employees from various generations working within the same organization) is becoming a frequent phenomenon (Winnicka-Wejs, 2020; Holian, 2015). At present, multigenerational teams can be considered to be an integral part of a social organizational architecture. Multigenerational teamwork is one of the effective tools for competitiveness and a source for a company's advantage. It should also be considered to be one of the components of Industry 4.0, where the development of technology is accompanied by an increasing level of robotization. Support for such processes requires the effective cooperation of employees from different generations.

Therefore, the question arises: how should multigenerational teams be effectively managed in the era of Industry 4.0? A hypothesis was put forward that generational intelligence is a determinant of the development of a 21st-century organization. This chapter reviews the literature that addresses the problem of multigenerationality in management and presents the advantages and disadvantages of managing a multigenerational team. Moreover, it explains the concept of Industry 4.0. The considerations are concluded with a presentation of the authors' model of the multi-faceted management of a multigenerational team.

Phenomenon of Employee Multigenerationality as a Challenge in Managing Modern Organization

Turbulence in the market environment, the intensification of globalization, the deepening interdependencies between business entities, and the emergence of integration processes under the conditions of market instability have become factors that imply difficult challenges for modern organizations. In order to survive, these entities have been forced to evolve, improve their ability to think strategically, and react quickly to emerging market threats. They are therefore forced to effectively manage their employee teams (Beazley et al., 2017).

DOI: 10.4324/9781003186373-14

In the current economic reality, there have been many changes in human resource management (Schubert & Andersson, 2015). New, young employees keep joining the team in the workplace; they will have to work with older people or even manage them. Such a diversity of employees affects the functioning of a workplace in which representatives of different generations are employed.

The phenomenon of multigenerationality is recognized as an essential element of the social architecture of modern organizations regardless of their size or type of business. Managing a multigenerational team is now recognized as one of the tools of competitiveness, innovation, and leadership. Lyons, Schweitzer, and Eddy (Lyons et al., 2015; Moczydłowska, 2016) express the view that multigenerationality remains a significant source of a lasting competitive advantage that largely determines the success of an organization. In recent years, the age structure of employees has radically changed. Representatives of different generations work together now; therefore, new challenges have emerged for contemporary organizations due to the fact that generational diversity has become an increasingly common phenomenon in contemporary business environments. According to different authors (Wang et al., 2019; Wang & Haggerty, 2009), the modern organization brings together the greatest diversity of generations as compared to any other moment in history. Many organizations have representatives of several generations working together; these employees are characterized by different life experiences, skills, and values. Economic history shows that technological revolutions tend to generate deep economic and social crises before a temporary state of equilibrium is reached (Huebner, 2005; Aspromourgos & Lodewijks, 2004).

Characteristics of Generations on Labor Market

Today, representatives of four generations meet in the labor market. These generations differ in lifestyles, communication methods, values, needs, and motivations. They also bring a unique compilation of skills, talents, knowledge, and competencies to an organization. They include Baby Boomers as well as Generations X, Y, and Z. Detailed information is shown in Table 14.1.

Table 14.1 Generations in labor market

Baby Boomers		*Generation X*		*Generation Y*		*Generation Z*	
Born during following periods							
1946	1964	1965	1976	1977	1990	1995	2018

Source: Żarczyńska-Dobiesz & Chomątowska (2014), p. 407

Baby Boomers are those employees born during the years of 1946–1964. The features that characterize this generation are their recognized authority and position in the labor market. Their patriarchal perception of reality was built on professional successes. Women are seen mainly as mothers and wives; it is the man who is the guarantor of financial security and stability (Holian, 2015; Wang & Haggerty, 2009). For people from this generation, it was common that fathers were permanently absent from the life of the family and that the model of raising children was based on the provision of resources and protection against potential threats. The role of work was most accurately characterized by the saying "no pain, no gain." This generation has great respect for authorities, adheres to established rules, and respects the hierarchical structure in the workplace (Gardner, 2009; Szpringer et al., 2014).

The term "Generation X" refers to people born during the years of 1965–1976 (Jurkiewicz, 2000; Wells et al., 2014). This generation was marked by the development of consumption in the 1980s and early 1990s. They lived through such phenomena as the manipulation of the political system, rapid political changes (the collapse of the USSR, fall of the Berlin Wall, and end of the Cold War), the popularization of the Internet, and the emergence of diseases of affluence. This generation is also often called the "MTV Generation" and "Generation Jones." These slogans refer to the desire to achieve professional success. People from this generation are most often involved in performing their professional duties; they are responsible and interested in the world. Due to the fact that they were born during the development boom, they remain great promoters of technology. Equally important are active outdoor recreation and meeting people. People from Generation X were born during times of dynamic change and do not need the Internet at all to spend their free time (Chomątowska & Smolbik-Jęczmień, 2013).

The name "Generation Y" is based on the preceding Generation X. People who belong to this generation were born during the demographic boom that took place in the 1980s and early 1990s (worldwide). In Poland, the timeframe of this generation is wider. It covers the years of 1980–1999 (Wong et al., 2017). These discrepancies are conditioned by the fact that the important factor that shapes Generation Y is the experience of capitalism and the rapid development of new technologies. In Poland, these events did not take place until after the fall of communism in 1989. It was then that two processes overlapped in our country: an economic transformation took place, and the Internet became widespread. This explains why the experiences of Poland's Generation Y are delayed as related to their peers living in Western Europe or the United States (Mazur-Wierzbicka, 2015).

People from Generation Y are multi-tasking employees who are engaged in their work and undertake new challenges. They value flexible work time and teamwork; they are interested in work that is satisfactory in terms of salary and their own professional development. These employees can be described as highly socialized individuals (Gadomska–Lila, 2015).

The next generation after Generation Y is Generation Z; in some sources, they are also referred to as "Generation C" (from "connected"—connected to the Internet). Literature sources lack an explicit agreement over the age of people from this generation. Usually, people included in Gen Z are those born after 1990 (and according to some sources, after 1995). This generation is composed of people who were born during the years 1991–2000. This means that a large proportion of people from this generation are still continuing their education. Some, however, have already become part of the staff of modern enterprises operating in the market (Tulgan, 2009). It is interesting that the people included in Generation Z have numerous features in common with the previous generation. Nevertheless, there is agreement among researchers that they are similar to Generation Y in some areas while being fundamentally different in many others (Knapp et al., 2017; Dolot, 2018; Half, 2015).

Generations in Labor Market During Era of Industry 4.0.

The progressing digitization of whole societies has dramatically changed production methods and purchasing structures. Technology connects numerous industries with society and affects the volume and quality of production as well as the functionality of products. What is characteristic of Industry 4.0 is the introduction of leading IT solutions in all areas of production, which allows for the creation of not only specific products but also entire related value chains (Vaidya et al., 2018).

The use of advanced ICT technologies allows for a more accurate adjustment of production to different customer expectations while maintaining low costs as well as high levels of quality and efficiency. New business models (including artificial intelligence or additive manufacturing) enhance industry transformation processes and change current business methods and market structures (Xu et al., 2018).

All generations functioning in contemporary organizations operate during the era of Industry 4.0. This concept assumes that, despite the advanced technologies, it is people with their unique abilities who are the most important component of an organization. However, all generations in the labor market will receive much more support than before thanks to the new solutions.

The work of different generations during the era of Industry 4.0 is no longer the future but the present; this is why contemporary employees' perception of the industry should change. Cyberbiological systems give rise to new challenges, connecting machines with people so that the latter are able to control the whole process.

Challenges in Managing Multigenerational Teams

Interest in age management continues to increase in the labor market. This phenomenon can now be considered to be part of the diversity management process as a permanently rooted element of human resource management. It focuses largely on the implementation of various activities that allow for the rational and effective use of human resources in workplaces by taking the needs of employees of different ages into account (Citkowski & Garwolińska, 2018).

By analyzing the socio-historical approach to such a problem, the term "generation" can be used to describe a cohort of people connected by certain experiences they had during their youth (mainly, important collective events). In addition, generational affiliation must be considered to be an objective attribute associated with being born during a specific period of time as well as a set of similar experiences and specific socio-historical determinants that make up the context of the further functioning in life. At the same time, the generational approach should not automatically amount to disregarding individual differences (Chomątowska & Smolbik-Jęczmień, 2013).

Employees representing different generations follow different values differ in their needs, lifestyles, approaches to duties, motivations, and manners of communicating (Katzenbach & Smith, 2001). However, it is this diversity that allows them to bring a unique composition of knowledge, skills, competencies, abilities, talents, etc., to an organization (Młodzi, 2011). The functioning of a multigenerational team may prove to be one of the main competitive advantages and a source of teamwork success. Therefore, a diverse team of employees in terms of age should be perceived as an opportunity rather than a threat, especially since a management's task is to use the strongest sides of a team's form of work organization (Klaffke, 2014).

Efficiently managed multigenerational teams can become an essential structural element of modern organizations, especially since enormous potential lies in the work of teams that are varied in terms of age. Nevertheless, its fulfillment is conditioned by numerous internal and external factors that have their sources in a company's environment (Yang & Matz-Costa, 2018).

The conscious design of age-diverse teams is becoming a huge reservoir of new ideas; this makes it possible to use the intellectual potential inherent in people, enables the implementation of innovative solutions, and promotes social integration. Working on a multigenerational team overcomes stereotypical ways of thinking. Employees of different ages share valuable knowledge, skills, abilities, and experiences (Hammermann et al., 2019).

The mutual exchange of attributes that are typical of a given employee's age builds effective cooperation in a multigenerational team, which has become an important structural element of modern organizations in the current economic reality (Pyka, 2009). There is great potential in the work of such teams; properly utilized, it can bring great benefits for organizations and team members. The process of the conscious creation of multigenerational teams in an enterprise favors the use of the intellectual potential that is inherent in one's employees (Gordon, 2018).

Managing a multigenerational team during the era of Industry 4.0 is a huge challenge. This will be especially difficult for the oldest generation to find themselves in the new reality; that is why cooperation with younger generations will become so important for them, as the younger employees will help them function in the face of a huge amount of available data and computational possibilities.

It should also be borne in mind that the functioning of teams with members of different ages may be accompanied by some difficulties. Employees from the Baby Boomer generation often see younger generations as employees who lack discipline. They believe that younger colleagues do not comply with the rules that apply in the workplace. On the other hand, employees representing Generations Z or Y are often seen as entitled and unable to effectively communicate with colleagues in a direct way due to the fact that they grew up in the era of digital technology (Kaczmarek & Krajnik, 2009).

Practice therefore shows that the reason for conflicts in the workplace is often older people's perception of young employees' approaches (Jurewicz, 2000). Employees from the oldest generation complain that younger employees are not responsible for their work, want to achieve goals by taking short-cuts, lack diligence and patience, demand awards and promotions without putting effort into their work, and make hasty decisions to leave the workplace (Savanevičienė et al., 2019).

In turn, younger employees perceive older workers as people who lack flexibility in action and are attached to rigid rules and practiced working methods to obsolete and outdated solutions (Heyns & Kerr, 2018). Employees from the oldest generation are usually unwilling to accept new changes and are very slow to become acquainted with the latest market developments (Oliveira & Cabral-Cardoso, 2017). In addition, they are not very willing to share their experience and knowledge, they rarely appreciate the competencies of younger employees, and they block their creative ideas (often cooling their enthusiasm) (Žukauskas, 2018). In many workplaces, a discriminatory approach to older workers can also be seen, and their potential is overlooked (Shelley André, 2018).

A lack of mutual trust can be considered to be a considerable problem in managing a multigenerational team. Also, a mutual lack of trust remains a problem. Representatives of different generations approach mutual cooperation with great caution and distrust (and often hostility and suspicion). Such problems in mutual relationships have various consequences. They usually hinder the day-to-day fulfillment of duties, deteriorating the quality of work, the efficiency of individual employees and entire teams, and (ultimately) the workplace as a whole (Wang & Fang, 2020).

Therefore, a great challenge is the proper and effective management of multigenerational teams as a category of diversity management. It is this diversity that allows for making use of the experience, knowledge, skills, and predispositions of a company's employees (Durska, 2009). The problem of age management due to the presence of different generations in the labor market has been a field for discussion for several years among theoreticians and practitioners of human resource management. However, they all agree that it is necessary to guarantee such working conditions that allow for the best possible use of human resources. This will strengthen the market position of an organization and lead it to achieving a competitive advantage (Earl & Taylor, 2015).

The management of a team that is diverse in terms of age should focus on the effective use of the skills, experiences, and competencies of generations socialized under different economic, historical, and social circumstances.

It often happens that management does not notice problems related to the functioning of employees differentiated in terms of age in one's workplace. Therefore, it is very important to develop a good working atmosphere that enables employees to get to know each other's needs. A friendly and positive atmosphere in the workplace means that the staff will be more open; they will also feel respected and appreciated. The role of management is to also identify the strengths and weaknesses of each generation as well as promote tolerance, integration, and respect. Only creating such conditions will make it possible to make full use of the employees' potential (Beazley et al., 2017; Bieling et al., 2015).

It is equally important to combat the stereotypes that affect the perception of employees from different generations. The appropriate management of a staff that is diverse in terms of age favors a greater motivation and commitment of one's employees; this improves their efficiency and helps a company acquire talented workers. It also favors the creation of innovative ideas and allows for overcoming established stereotypes (Chawla et al., 2017).

It is also important to make multigenerational teams aware of the huge possibilities provided by the wide use of mobile communications for transmitting data from devices, enabling them to connect devices through the Internet of Things (IoT). It is necessary to realize that modern staff functions in the era of the progressive automation of production where robots are becoming increasingly popular (i.e., during the period of Industry 4.0).

All generations functioning together in a workplace should develop a common system of mutual cooperation. Older generations in particular need to learn how to function in a reality that is based on the automatic collection and processing of large data sets, either from devices or directly from people (Graystone, 2019).

The increase in the amount of available data and computational capabilities has become the modern reality. As a result, it has become possible to better manage a company's resources, plan the production, or manage an entire product life cycle. In such an environment, effective cooperation between generations provides an opportunity to deepen cooperation with suppliers as well as better respond to customer needs (Kim, 2018; Desai & Lele, 2017).

Generational Intelligence as Determinant of Development of 21st-Century Organization

Proper communication becomes a key issue in managing a multigenerational team. The aging process of societies may cause the intensification of workplace conflicts and may result from deep ambivalence in the public and private spheres. These problems are already present in almost every organization; therefore, the management of each company should strive to solve them. The development of generational intelligence will certainly be an effective method.

This term was first used by Biggs and Lowenstein (2011). This term has also spread in the Polish literature on the subject. The definition by Moczydłowska should be quoted here, as it describes generational intelligence as the ability to reflect and act as a derivative of understanding one's own and other people's "course of life, family, and social stories placed in their social and cultural contexts" (Warwas, 2016).

The phenomenon of generational intelligence must also be combined with the desire to reduce intergenerational conflicts and disputes as well as the necessity to strengthen the value of multigenerational capital. These activities correspond to the needs of the concept of a learning organization. According to its assumptions, an intergenerational transfer of knowledge, skills, competencies, and abilities is necessary for the proper functioning of an enterprise (Warwas, 2016).

In response to these needs and the emerging demographic and economic conditions, the phenomenon of generational intelligence has developed. This phenomenon assumes the skillful use of existing human capital under the conditions of the trend of an increasing average age of employees (Moczydłowska, 2016).

Generational intelligence sets itself the task of creating lasting relationships between generations that function against the background of changing sociodemographic trends. The phenomenon of generational intelligence allows for an analysis of contemporary social problems and their potential solutions. It can therefore be considered as an effective method of studying the interactions between younger and older generations (Warwas, 2016). Generational intelligence is usually measured by the level of the tasks performed, the quality and speed of their performance, and the level of learning ability (Stefaniak et al., 2015; Haapala et al., 2014).

Managing Multigenerational Team in Model Approach

In addition to conducting an analysis of generations in the labor market, the authors of the present chapter have attempted to construct a model of a multigenerational team functioning in a modern organization. These considerations are illustrated in Figure 14.1, where it is pointed out that a multigenerational team composed of different generations with a unique personality and temperamental traits creates the generational intelligence of a modern enterprise. This intelligence bears the hallmarks of linguistic, emotional, social, strategic, and interpersonal intelligence and leads to strengthening the market position of a company and improving its competitiveness.

It has been shown that a multigenerational team can function in the era of Industry 4.0. The new reality marks the final end of Taylor's concept of hierarchical management. Teamwork, autonomy, low-level decision-making, and innovation are becoming the most important values. The currently used solutions provide flexibility that has not been achieved until now in adapting to customer expectations (and thus, also have an advantage over the competition).

Conclusions

In the contemporary turbulent labor market, representatives of different generations with different lifestyles, values, communication methods, motivations, and needs are simultaneously present. Proper communication and, at the same time, understanding the distinctness of each generation is undoubtedly a key issue in managing a modern enterprise. Employees who are representatives of different generations have different values, needs, goals, approaches to work, methods of communication, etc. This diversity means that they bring a unique set of knowledge, skills, abilities, competencies, talents, etc. to the workplace.

Multigenerational teams have now become an inseparable element of the functioning of a modern organization in the era of Industry 4.0, where new technologies and processes are being introduced. The use of generational intelligence is now recognized as an effective competitiveness tool as well as a source of a company's advantage.

Recommendations

Multigenerational teams can surely increase the value of any organization and, due to the exchange of experiences, reduce the level of stress

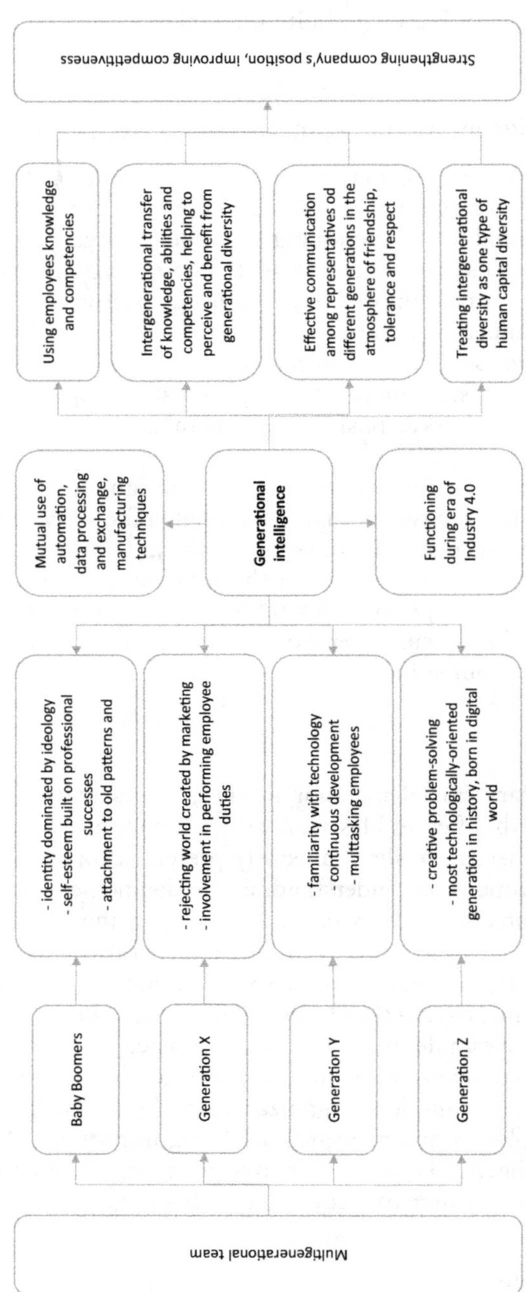

Figure 14.1 Model of multigenerational team functioning in a modern organization

Source: authors' elaboration

and ease conflicts and tensions. This is why it is so important to pass on some recommendations for staffs that manage a multigenerational team:

1. A multigenerational team should be perceived as a source of productivity that will contribute to the success of the entire organization.
2. The different expectations and needs of the employees should be considered while managing a multigenerational team.
3. In order to reduce the level of stress, it is important to make the employees aware of their multigenerational differences. Therefore, mutual communication and organizing cyclical meetings that allow for the discussion of potential problems are so essential. Such actions would help to mutually assimilate and effectively make use of experienced resources.
4. Younger employees could familiarize older employees with new applications and technologies while eliciting professional experience from them.
5. Additionally, all employees should be responsible for decisions and duties.

These recommendations can be helpful for company executives who are facing the challenge of managing a multigenerational team. These rules can be successfully introduced to the US market, as it is not uncommon to find teams of different ages in companies. Americans enter the labor market quite early; however, older people are still professionally active. The rules presented in the chapter could therefore be applied to the management of US companies, both those of the sizes of Pepsico or Procter and Gamble and smaller ones. This could inspire further research.

References

Aspromourgos, T., & Lodewijks, J. (Eds.). (2004). *History and Political Economy: Essays in Honour of P.D. Groenewegan*. Routledge.

Beazley, A., Ball, C., & Vernon, K. (2017). Workplace age diversity: The employers' perspectives. In M. Flynn, Y. Li, & A. Chiva (Eds.), *Managing the ageing workforce in the east and the west. (The changing context of managing people)* (pp. 225–247). Emerald Publishing Limited.

Bieling, G., Stock, R. M., & Dorozalla, F. (2015). Coping with demographic change in job markets: How age diversity management contributes to organisational performance. *German Journal of Human Resource Management*, *29*(1), 5–30.

Biggs, S., & Lowenstein, A. (2011). *Generational intelligence: A critical approach to age relations*. Routledge.

Chawla, D., Dokadia, A., & Rai, S. (2017). Multigenerational differences in career preferences, reward preferences and work engagement among Indian employees. *Global Business Review*, *18*(1), 181–197.

Chomątowska, B., & Smolbik-Jęczmień, A. (2013). Zespoły wielopokoleniowe wyzwaniem dla współczesnego organizatora pracy w warunkach nowej gospodarki. Zeszyty Naukowe Uniwersytetu Szczecińskiego. *Ekonomiczne Problemy Usług, Europejska przestrzeń komunikacji elektronicznej*, *105*(2), 193–202.

Citkowski, M., & Garwolińska, M. (2018). Strategia zarządzania wiekiem w przedsiębiorstwie jako odpowiedź na wyzwania rynku pracy. Przedsiębiorczość i Zarządzanie, Wyzwania w zarządzaniu zasobami ludzkimi we współczesnych organizacjach. *Od teorii do praktyki*, *19*(1/8), 461–474.

Desai, S. P., & Lele, V. (2017). Correlating internet, social networks and workplace-a case of generation Z students. *Journal of Commerce and Management Thought*, *8*(4), 802.

Dolot, A. (2018). The characteristics of Generation Z. *E-mentor*, *74*(2), 44–50.

Durska, M. (2009). Zarządzanie różnorodnością: kluczowe pojęcia. *Kobieta i Biznes*, *1–4*, 8.

Earl, C., & Taylor, P. (2015). Is workplace flexibility good policy? Evaluating the efficacy of age management strategies for older women workers. *Work, Aging and Retirement*, *1*(2), 214–226.

Gadomska–Lila, K. (2015). Pokolenie Y wyzwaniem dla zarządzania zasobami ludzkimi (Generation Y as a challenge to human resource management). *Zarządzanie zasobami ludzkimi*, *1*(102), 25–39.

Gardner, H. (2009). *Inteligencje wielorakie: nowe horyzonty w teorii i praktyce*. MT Biznes.

Gordon, P. A. (2018). Age diversity in the workplace. In *Diversity and inclusion in the global workplace* (pp. 31–47). Palgrave Macmillan.

Graystone, R. (2019). How to build a positive, multigenerational workforce. *JONA: The Journal of Nursing Administration*, *49*(1), 4–5.

Haapala, H. L., Hirvensalo, M. H., Laine, K., Laakso, L., Hakonen, H., Lintunen, T., & Tammelin, T. H. (2014). Adolescents' physical activity at recess and actions to promote a physically active school day in four Finnish schools. *Health Education Research*, *29*(5), 840–852.

Half, R. (2015). *Get ready for generation Z*. Maclean's.

Hammermann, A., Niendorf, M., & Schmidt, J. (2019). *Age diversity and innovation: Do mixed teams of" old and experienced" and" young and restless" employees foster companies' innovativeness?* Institute for Employment Research 4/2019. IAB-Discussion Paper.

Heyns, M. M., & Kerr, M. D. (2018). Generational differences in workplace motivation. *SA Journal of Human Resource Management*, *16*, 10.

Holian, R. (2015). Work, career, age and life-stage: Assumptions and preferences of a multigenerational workforce. *Labour & Industry: A Journal of the Social and Economic Relations of Work*, *25*(4), 278–292.

Huebner J. (2005). A possible declining trend for worldwide innovation. *Technological Forecasting and Social Change*, *72*(8), 28–29.

Jurkiewicz, C. L. (2000). Generation X and the public employee. *Public Personnel Management*, *29*(1), 55–74.

Kaczmarek, P., Krajnik, A., Morawska-Witkowska, A., Remisko, B. R., & Wolsa, M. (2009). *Firma = różnorodność. Zrozumienie, poszanowanie, zarządzanie*. Forum Odpowiedzialnego Biznesu.

Katzenbach, J. R., & Smith, D. K. (2001). *Siła zespołów. Wpływ pracy zespołowej na efektywność organizacji.* Merlin.

Kim, S. (2018). Managing millennials' personal use of technology at work. *Business Horizons, 61*(2), 261–270.

Klaffke, M. (2014). *Generationen-Management.* Konzepte, Instrumente, Good-Practise-Ansätze, Hrsg, Springer Gabler.

Knapp, C. A., Weber, C., & Moellenkamp, S. (2017). Challenges and strategies for incorporating Generation Z into the workplace. *Corporate Real Estate Journal, 7*(2), 137–148.

Lowstein, A. E., (2011), Early care, education, and child development. *Annual Review of Psychology 62*(1), 483–500.

Lyons, S. T., Schweitzer, L., & Eddy, S. W. (2015). How have careers changed? An investigation of changing career patterns across four generations. *Journal of Managerial Psychology, 30*, 8–21.

Mazur-Wierzbicka, E. (2015). Kompetencje pokolenia Y—wybrane aspekty. *Studia i Prace WNEiZ US, 39*(3), 307–320.

Młodzi. (2011). Młodzi 2011 | Narodowe Centrum Kultury (nck.pl).

Moczydłowska, J. M. (2016). Zarządzanie relacjami międzypokoleniowymi jako cecha przedsiębiorstw inteligentnych. Organizacja inteligentna. *Perspektywa zasobów ludzkich*, 227–238.

Oliveira, E., & Cabral-Cardoso, C. (2017). Older workers' representation and age-based stereotype threats in the workplace. *Journal of Managerial Psychology, 32*(3), 254–268.

Pyka, M. (2009). Pomiędzy "Homo Sovieticus" A "Homo Aemulnns". Uwagi o idei solidarności i potrzebie filozofii pracy w społeczeństwie rynkowym. *Warszawskie Studia Teologiczne, 22*, 103–114.

Savanevičienė, A., Stankevičiūtė, Ž., Navickas, V., Grėbliūnaitė, M., & Okręglicka, M. (2019). Crucial work environment factors for different generations' employee: Organisation fit. *Polish Journal of Management Studies, 19*, 364–375.

Schubert, T. & Andersson, M. (2015). Old is gold? The effects of employee age on innovation and the moderating effects of employment turnover. *Economics of Innovation and New Technology, Taylor & Francis Journals, 24*(1–2), 95–113.

Shelley André, R. N. (2018). Embracing generational diversity: Reducing and managing workplace conflict. *ORNAC Journal, 36*(4), 13.

Smolik-Jęczmień, A., (2013). Podejście do pracy i kariery zawodowej wśród przedstawicieli generacji X i Y - podobieństwa i różnice. *Nauki o Zarządzaniu, Uniwersytet Ekonomiczny we Wrocławi, 1*(14), 89–97.

Stefaniak, A., Bilewicz, M., & Winiewski, M. (Eds.). (2015). *Uprzedzenia w Polsce.* Wydawnictwo Liberi Libri.

Szpringer, M., Kopik, A., & Formella, Z. (2014). Multiple intelligences and minds for the future in a child's education. *Journal Plus Education*, 350–359.

Tulgan, B. (2009). *Not everyone gets a trophy: How to manage generation Y.* John Wiley & Sons.

Vaidya, S., Ambad, P., & Bhosle, S. (2018). Industry 4.0—a glimpse. *Procedia Manufacturing, 20*, 233–238.

Wang, M., & Fang, Y. (2020). Age diversity in the workplace: Facilitating opportunities with organizational practices. *Public Policy & Aging Report, 30*(3), 119–123.

Wang, W. T., Wang, Y. S., & Chang, W. T. (2019). Investigating the effects of psychological empowerment and interpersonal conflicts on employees' knowledge sharing intentions. *Journal of Knowledge Management*, *23*(6), 1039–1076.

Wang, Y., & Haggerty, N. (2009). Knowledge transfer in virtual settings: The role of individual virtual competency. *Information Systems Journal*, *19*(6), 571–593.

Warwas, I. (2016). Zarządzanie generacjami na wewnętrznym rynku pracy. In I. Warwas & A. Rogozińska-Pawełczyk (Eds.), *Zarządzanie zasobami ludzkimi w nowoczesnej organizacji. Aspekty organizacyjne i psychologiczne*. Wydawnictwo Uniwersytetu Łódzkiego.

Wells, J. R., Środa, J., & Bukowski, A. (2014). *Inteligencja strategiczna: jak stworzyć mądrą firmę*. Dom Wydawniczy REBIS.

Winnicka-Wejs, A. (2020). Deficits and potentials: How risk involving generational characteristics can be reduced thanks to human capital multigenerationality. *Human Resource Management/Zarzadzanie Zasobami Ludzkimi*, *133*(2), 41–56.

Wong, I. A., Wan, Y. K. P., & Gao, J. H. (2017). How to attract and retain Generation Y employees? An exploration of career choice and the meaning of work. *Tourism Management Perspectives*, *23*, 140–150.

Xu, L. D., Xu, E. L., & Li, L. (2018). Industry 4.0: State of the art and future trends. *International Journal of Production Research*, *56*(8), 2941–2962.

Yang, J., & Matz-Costa, C. (2018). Age diversity in the workplace: The effect of relational age within supervisor—employee dyads on employees' work engagement. *The International Journal of Aging and Human Development*, *87*(2), 156–183.

Żarczyńska-Dobiesz, A. & Chomątowska B. (2014). *Pokolenie, "Z" na rynku pracy—wyzwania dla zarządzania zasobami ludzkimi*. Wydawnictwo Uniwersytetu Ekonomicznego we Wrocławiu.

Žukauskas, P., Vveinhardt, J., & Andriukaitienė, R. (2018). Sociodemographic indicators: Employee attitude. *Management Culture and Corporate Social Responsibility*, *249*, 351–429.

15 Progress in Implementation of Fourth and Fifth Industrial Revolutions and Artificial Intelligence

Barbara Siuta-Tokarska and Agnieszka Thier

Introduction

Industrialization is an important factor in socio-economic development and civilization progress, as it determines the structural changes in an economy as well as social life. Industrialization and technical progress in the industry determine the increase in the quality of life not only by providing more modern and relatively cheap products but also as a result of its impact on the development of services. Along with increases in the level of processing in the industry, the share of the service sector in the employment of an economy as well as in the generation of national income (GDP) increases. This has resulted in a new look at the structure of the economy and the role of industry in its development, which resulted in the concept of the emergence of a *post-industrial* society and the advent of the post-industrial era as well as the *servicization* phase; that is, the superiority of services over the other economic sectors. However, it turned out that the so-called *deindustrialization* cannot go too far, as many service providers are unable to operate without the supply of new industrial products. Hence, the birth *of a strong reindustrialization* slogan has been proclaimed by European Union agencies (among others) and reflected in the economic development programs of many countries, including the United States. Industry is not so much subject to internal development mechanisms—stimulated by the economic policies of states and international organizations; however, in its global development tendency, one can distinguish phases of revolutionary changes in their nature. Thus, the global economy is currently entering the fourth stage of the industrial revolution (i.e., the digital revolution); in the most developed countries, there are already manifestations of the Fifth Industrial Revolution (i.e., the dissemination of technologies based on artificial intelligence and the synergistic cooperation between people and machines).

The purpose of this chapter is to present the essence, manifestations, and economic effects as well as the social effects of the Fourth Industrial

DOI: 10.4324/9781003186373-15

Revolution (which are still difficult to assess against the background of previous stages of industry development) as well as to analyze its first practical manifestations in selected countries (including Poland). In addition, the authors discuss the issue of the challenges related to artificial intelligence.

Stages of Industrial Revolution

The term *industrial revolution* was introduced in 1884 by British historian Arnold J. Toynbee, who traced the first revolution back to 1760. He associated this with the emergence of the factory industry as a result of the introduction of steam propulsion in machinery in England and Scotland (Blung, 2000, p. 35). However, the symbolic year was considered to be 1784, when the James Watt steam engine was installed in a weaving workshop. In turn, the Second Industrial Revolution occurred at the cusp of the 19th and 20th centuries due to the emergence of electricity and the invention of several industrial products such as the internal combustion engine, electric bulb, telephone, radio, and camera as well as the introduction of line production and the subsequent mass production. The first was a meat-processing line in Cincinnati, OH (launched in 1870), and the most famous one was Henry Ford's car production line (launched in 1913). Whereas the Third Industrial (information) Revolution marks the use of computers and the Internet in the latter half of the 20th century; thus, full control of the production process and flexible production systems as well as the use of robots. A symbolic year is 1969, when the Modican 084 programmable logic controller (PLC) was introduced, which began the era of industrial automation. In addition to the classic factors of production (land, capital, and labor), the key role is then played by *knowledge (technology)*.

The Third Industrial Revolution marked a big step in modernizing industry due to IT and the use of industrial automation; however, it caused the economy to enter a *post-industrial* phase at the same time. Deindustrialization would be a stronger term. In Poland and the Eastern European countries, this process was also intensified due to political changes. Contrary to the name, deindustrialization does not mean the decline of industry—although decapitalized and obsolete industrial plants are, of course, closed down—but a decrease in its relative importance in the employment structure of the economy and in creating national income. The share of industry in the creation of GDP currently reaches about 22–23% in the United States and the European Union (including Poland), while the share of services is approximately 75–77% (in Poland, this is about 67%, and in China—only 42%). Therefore, the structural changes are better reflected not so much in the decline in the importance of industry as they are in exposing the development of services, that is, the process of the servicization of the economy (Kiełczewski, 2012).

However, the excessive dominance of finance over the real sphere as well as the service sector over industry and construction lead to economic disparities. As a result of the criticism of these disparities, the idea of *reindustrialization* has been taken up. In particular, the European Union inspired the "Investments for Europe" program in 2014 and established the European Fund for Strategic Investments, which was followed by the *circular economy* program as a higher stage of *sustainable development*. In Poland, on the other hand, the "Strategy for responsible development" program (also referred to as the reindustrialization program) was launched (Council of Ministers, 2016). This program has ambitious goals; however, there is a lack of financial and organizational resources for its implementation according to many critics (even up to 2030).

These programs not only emphasize concern for jobs and the fight against unemployment but also accent the importance of providing modern products and technologies for services (thus maintaining a balance between industrial production and service provision). All of this argues in favor of the emergence of a new stage of socio-economic development and civilization progress, which is called the Fourth Industrial Revolution (i.e., the digital or technological revolution). Its synonym is the term *fourth-generation industry* or Industry 4.0. Therefore, executives (including those at lower levels of the economy) should be aware of the directions of development of the Fourth Industrial Revolution and carry out the assessments of technical progress in this context (in industrial as well as service enterprises) in order to benefit from new solutions and implementation patterns as well as from the assistance provided for in the EU and government programs in this area.

To sum up, the term "Industrie 4.0" was initiated by the German Federal Government during the Hannover Messe (Hannover Fair) in 2011 (Sawhney et al., 2020; Tortorella, Mac Cawley Vergara et al., 2020; Tortorella, Pradhan et al., 2020). Thus, Industry 4.0, which is also referred to as *Smart Industry* or integrated industry, combines machines, processes, and products info intelligent networks (Szász et al., 2021). Some publications reveal more than 100 definitions of Industry 4.0 (Moeuf et al., 2017a, 2017b; Tobon Valencia et al., 2018). They are different but similar in reality.

Manifestations and Implementation of Fourth Industrial Revolution

The Fourth Industrial Revolution is the result of the digitization of production processes as well as the growing role of information and the use of cyber-physical systems. It consists of creating a networked social system that connects people, technical infrastructure, and production and service devices throughout the entire value-added chain (from order placement to customer receipt). Therefore, the digitization of industry means increasing the amount of information and accelerating its transmission, and *online* communication is increasingly taking place not between man and machine but between two machines (*machine-to-machine*—the so-called M2M), which contributes to limiting the role of human work (Olender-Skorek, 2017). The universality, speed, and quality of information based on information and communication technologies (ICT) enable the creation of intelligent factories that are able to adapt and optimize the use of resources. An important characteristic is also focusing on individual selection features and integrating the customer with the manufacturer—due to digital access (i.e., *personalizing products*). The symbol of the Fourth Industrial Revolution is the smartphone as well as digital cyber-physical systems consisting of intelligent sensors (among other things) that also react to human presence, software, and communication systems known as the *5C architecture* (that is, *connection, conversion, computerization, cognition, and configuration*) along with advanced robots, 3D printers, autonomous vehicles, and drones. Novel digital technologies include "cloud computing," big data applications, mobile technologies, and social media.

The use of the concept of the Fourth Industrial Revolution was initiated in Germany in 2010–2011. A *Platform Industrie 4.0* working group was then set up as a contact center for industry, business, and science. Thus, Industry 4.0 (also referred to as Smart Industry or integrated industry) combines machines and devices, processes, and products together into intelligent networks.

The third revolution consisted of the automation of individual machines and production processes; this stage of enterprise development is not yet completed even in developed countries (including Poland). The Fourth Industrial Revolution, on the other hand, turns out to be a comprehensive digital transformation of all fixed assets and their integration with partners who co-create the value-added chain within the digital ecosystem. Such a system works efficiently and is generally cheap, for example, the costs of data storage have drastically decreased (from 10,000 to 0.03 USD for 1 GB/year). However, these solutions can only be used provided that the staff and infrastructure are properly prepared, which is capital-intensive. This is a reminder of how companies once implemented improvements in occupational health and safety and then again in the 1970s with environmental protection equipment: it started with incurring economic outlays (in extreme cases; e.g., in the energy sector, it consumed up to 30–40% of investment outlays), which later resulted in increased work efficiency, waste management, improved competitiveness on the market, etc. Initially, the processes of change in the industry took place automatically; however, the Third Industrial Revolution was significantly influenced by the state. Currently, the role of the state in this matter is increasing. In particular, the development of electronic services used in all branches of the economy and in everyday life is accelerating, which creates a digital society.

The basic elements of the Fourth Industrial Revolution are as follows:

- *The Industrial Internet of Things—IoT*, that is, a way of collecting data by current, temperature, pressure, and noise sensors and storing them in the cloud;
- *Cyber-physical system—CPS*, that is, a system in which the physical world connects to the virtual world by using sensors and executive modules; information about reality is processed using the mathematical reflection of physical objects. The CPS explains the environment better; due to a network of sensors, it enables the monitoring of a patient's vital signs or, in the automotive industry, allows us to avoid collisions in traffic. In production, cyber-physical production systems (CPPSs) are used (Zaborovsky et al., 2015);
- *Cloud computing* is a technology for storing data in a virtual space (instead of company computers) with access from any place and any time due to the Internet (in a similar way as using an electronic e-mail box or social networking sites);
- *Big data*—rich data sets with high variability and diversity, whose processing and analysis is difficult but provides new knowledge;
- *3D printing*—this is the process of producing physical three-dimensional objects according to a computer model (initially, prototypes);
- Machine-to-machine communication (Moeuf et al., 2017a, 2017b).

Due to IT and network connections, the number of holding companies and capital groups as well as industrial complexes and clusters is growing, which means that the cooperation of enterprises becomes more efficient and reliable and transaction costs significantly lower under certain conditions of digitization. New forms of market models are emerging (e.g., the *sparing economy*) not only among households but also enterprises or new types—*strategic alliances*—(corporations that compete on the market but cooperate in the design of new products [for example, electric cars]). Digitization means that every product can contain digital information that exchanges with other products and with the environment during the production process (without human intervention). Due to this, intelligent manufacturing devices are autonomously included in subsequent phases of the production process. Production is personalized; that is, the customer becomes an active participant in the design of the product or service, choosing the configuration from a catalog of available options. Intellectual design and the implementation of orders are facilitated by incremental manufacturing technologies (e.g., 3D printers), among others.

In turn, the Industrial Internet of Things is a dynamic and global network of physical objects, systems, platforms, and applications that are capable of communicating and sharing intelligence with each other as well as with the external environment and people. A data set created in this way is used to monitor the production process of an entire factory and its visualization (a virtual equivalent of the real world) as well as for the calculations and analyses for making decisions and on-line control (in real time). This type of monitoring allows for the prevention of malfunctions and failures, among other things. For remote control, operators can also connect to the "cloud" by using tablets and smartphones. Generally, IoT enables the automation of processes throughout a factory, thus optimizing control and increasing energy efficiency, reliability, and safety. As a result, there is a rapid increase in labor productivity and production volume (practically impossible to achieve by traditional methods).

New technologies, computer hardware, and other devices are constantly emerging in this field (as are the types of networks). For example, the project of Swiss company Stoxx Limited based in Zug (which introduced the first stock exchange index administered by artificial intelligence in 2018) stands out from the latest developments in digitization. This is a method of processing large data sets that takes many factors into account (including social ones [e.g., the aging of a population]) much better than the DAX and Staxx Europe 600 indices used so far. This is expected to significantly improve the forecasting in the field of the stock market investment. In turn, Motteo Andretto announced the creation of a similar index to track changes in cryptocurrency rates. Due to the new technique, the DAX and Stoxx Europa 600 indices (prepared for the German Stock Exchange in Frankfurt am Main and based on share prices of 30 companies and a large number of other parameters) are calculated every second using the XETRA system. On the other hand, the Taiwan-based Han Hai Precision Industry Co. (known under the trade name of Foxconn) and 13 large factories in China and many branches in other countries (with more than 1 million employees and 40% of the global electronics, including products for the giant Apple) streamlines production by transforming plants into smart factories, where the productivity will increase several-fold. In addition, Faxconn Industrial Internet is a well-known brand of industrial robots and cloud-computing services. With 25 cosmetics factories (including the Warsaw Plant branch in Kanie, near Warsaw), French group L'Oreal is also a pioneer in implementing the Industry 4.0 concept. This group standardizes and unifies production, delivery, and storage processes using the FlexNet (Apriso) system with the Quality Control module. All tasks are automated and monitored on a regular basis at the level of a serial employee. Another good example is Slovakia, where a lot of intelligent technologies have been introduced by multinational companies from abroad (Závadská & Závadský, 2018).

When compared to the other European countries, Poland is worse in the ranking of implementing the fourth-generation industry than in the comparisons of socio-economic development indicators, as its level of computerization is lower than the EU average. Weak staff and the lack of access to capital as well as low labor costs mean that only 15% of larger factories are automated and 75% are partially automated. However, there are already clear signs of new trends. First of all, the market is developing for the production and service of 3D printers and their supply with appropriate materials. "Zortrax" S.A. (based in Olsztyn) is a well-known printer manufacturer. Some household appliances, toys, and house components are already being manufactured with the use of these printers. New materials are constantly created (such as graphene). Industrial plants are created that are geared toward smaller orders and tailored to individual demand and serving niche customers (Michałowski, 2018). The cooperation of the Digital Poland Foundation and the Information Processing Center at the National Research Institute in the programming of artificial intelligence deserves to be emphasized. This center creates intelligent IT systems using SI technology for the research sector and monitors the development of this technology in science and business (Sztuczna inteligencja, n.d.). In Poland, a humanoid NOA robot was built (among other things).

To sum up, the progress in digitization in Poland is already visible; however, entrepreneurs are not always aware or convinced of the inevitability of these processes nor the benefits of creating links and networks of enterprises (including holdings and clusters). In particular, this applies to numerous family businesses whose managers traditionally prefer the multigenerational survival of a company over immediate profits—and rightly so. They are often too cautious regarding the latest trends of economic development. Therefore, the easiest way to meet the challenges of the Fourth Industrial Revolution is to combine the computerization of a company with joining an enterprise network and then developing a program for introducing 3D printers and robots and automating selected production and service processes. As parts of clusters, there are opportunities for the more-intensive cooperation of independent entities, including individual specialists. In holding companies, however, the existing legal bases and digitization facilitate the integration of the parent and daughter companies.

The Essence of Fifth Industrial Revolution

The Fifth Industrial Revolution means that, in the near future, the digital revolution will cover all of the economic and social activities of the population as a whole. In its definition, attention is drawn to several characteristic features and issues that are not well-recognized as of yet (Furmanek, 2018; Tworóg & Mieczkowski, 2019; Iwano et al., 2016; Nahawandi, 2019):

- It will be an era of robots cooperating with each other and with people. Robots are not only widely used in industry but also in transport, agriculture, medicine, science and education, home services, security services, recreation, and entertainment. The combination of human activity and creativity with the speed, productivity, and consistency of robotic activities will embrace many more people with new trends. A synergy between people and robots will occur as a result of connecting the human brain with a computer by using a special implant. This will make it possible to control a smartphone with thoughts, which will especially serve patients with paresis, for example;

- The fifth revolution means combining 5G communication technology with artificial intelligence, which will fully develop the Internet of Things, automation, and smart factories. Life in a cyber-environment will combine reality with virtuality in a way that has never been known before;
- The new model of interpersonal and social relationships will focus more heavily on man (human-centric society) due to the use of artificial intelligence, among other things. Contrary to expectations, this will not make Industry 4.0 (or actually, digital manufacturing) transition into the fifth generation of industry but into the Digital Society (i.e., Society 5.0).

The 5G system (that is, the fifth-generation mobile technology for the remote and wireless transmission of the information at previously unimaginable data download speeds) will ensure mass connectivity between multiple devices. This will spread and multiply the benefits of the Fourth Industrial Revolution and also enable the unlimited control of people by government institutions and services as well as conglomerates, for example. The 4G system was introduced in 2010 and used in the largest companies in the world (mainly from the United States, such as Google, Amazon, Apple, and Facebook, but also a few from Europe, such as Swedish Ericsson and Finnish Nokia [albeit on a smaller scale]). The 5G system is to be launched in 2020; so far, it is very advanced in the United States, China, Japan, and South Korea (where the network is already in six cities). By 2025, the European Union wants to create legal and financial conditions for the transition of enterprises to a fully digital production system. Thus, enterprises in Poland must not only prepare for these new technologies but also for the resulting social changes in their immediate surroundings and among their employees. Therefore, it is worth raising these issues now by inserting the appropriate points in selected training programs and on other occasions.

Challenges of Artificial Intelligence

Progress in implementing the digital revolution puts artificial intelligence and its applications at the forefront. In an encyclopedic approach, *artificial intelligence* means IT departments must create models of intelligent behavior by using computer programs and, in a broader sense, the field of knowledge covering fuzzy logic, neural networks, robotics, and other departments. The goal of such ventures is the creation of computer programs and the construction of machines and other devices that are capable of performing selected functions of the human mind, including the creation of self-learning programs and structures. The key areas for the development of artificial intelligence are as follows:

- Image recognition and processing technologies;
- Speech processing and collection technologies;
- Automatic robots and vehicles;
- Virtual assistants;
- Machine learning, that is, machines can independently draw conclusions and take action.

The GPT-2 language model (multi-scale: 1.5 billion parameters from 8 million websites!) can be used as an example of a well-developed program. It can write press reports, analyses, summaries, translations from other languages, answer questions about a given text, and impersonate other well-known people and write biased comments. The last two examples raise concerns about the effects of the development of artificial intelligence. Threats are particularly associated with learning machines. Although computers have no consciousness and artificial intelligence is based on algorithms, power can be gained by those who create and construct these algorithms. The computer not only wins when pitted against chess masters but also against poker masters (who cheat), which again becomes a negative argument. However, biocybernetic Prof. Ryszard Tadeusiewicz is of the opinion that robots will not take over because computers lack consciousness, and robots do not have such a need (because they are built by man) (Tadeusiewicz, 2007). Therefore, Stephen Hawking and Elon Musk (the best-known critics of the further development of artificial intelligence) are incorrect in their assessment that robots threaten civilization (E. Musk, Mars rocket maker: "It is a greater threat than a nuclear weapon.").

Culture and art in the times of artificial intelligence intensifies the issue of the consequences of the mass copying of works of art and natural landscapes as well as the dissemination of them without space and time restrictions. It is easy to register anything, digitize it, and (what is controversial) modify it. Similarly, artificial intelligence will begin to compete with live music creators. The transformation of culture through artificial intelligence technology is, therefore, also becoming an increasingly acute social problem. For example, text editors currently improve grammatical errors and smooth one's style, but artificial intelligence in management and business procedures becomes impersonal and limits the role of an active person (especially when machines can be creative). Despite some fears that are related to artificial intelligence, computer programs are becoming indispensable, which means that human participation in creating culture clearly decreases. The effects of this phenomenon are difficult to predict. Consequently, an entrepreneur agreeing to the inevitable dissemination of robots and other devices with artificial intelligence (regardless of the magnitude of decisions) must place an even greater emphasis on the role of human and social capital in the enterprise management system. In addition, it can be anticipated that artificial intelligence devices will be more widely introduced by enterprises that have already implemented other digital technologies, treating this as the next stage of digitization. This may result in an increase in the competitiveness gap between those enterprises that are advanced in 4.0 technologies and those companies that are less innovative.

Economic and Social Effects of Fourth Industrial Revolution

The discussed processes of digitization and automation have already brought effects in terms of shortening production times as well as reducing the costs and increasing the sales values in enterprises to the amount of several percentage points per year (in a few years, this number could even increase to 30–60%). These economic effects are obtained due to the following factors (Schwab, 2017; Götz & Gracel, 2017):

- A new management model based on greater IT and analytical potential as well as flexible organizational structures and teams (modeled on previously known engineering/value analysis teams, for example);
- More accurate and faster market recognition (a product's time-to-market can be reduced by 20–50%);
- Precisely meeting the expectations of individual online customers at every stage of a product's design and implementation;
- Designing and producing products so that they can be easily reused in accordance with the concept of a circular economy;
- Automating analytical and computational works (which increases their efficiency by approx. 50%);
- Developing individual and niche products and creating new business models;
- Better use of resources and avoidance of machine downtime as well as a reduction of inventory costs by 20–50%;
- The availability of cheap public services in the "cloud" and reducing market entry barriers;
- Creating two-speed IT and analytical systems: flexibly "for now" and for a longer time;
- Owing to the use of advanced technologies and devices, many possibilities have been offered for design and innovation (Laudante, 2017).

It is worth quoting a fairly simple example that takes on new features and functions. Well, virtual reality allows you to surround yourself with three-dimensional space—after putting on the goggles—that is, the artificial world, or only its virtual elements. This makes it even easier to buy furniture because, due to the smartphone, we can see this piece of furniture in 3D in our rooms. In Western Europe and the United States, it is more and more often preferred for a product not to be cheap but to be quickly available and adapted ("printed") to the needs of individual customers. This means a smaller role of labor costs as well as economies of scale resulting in a reduction of unit production costs. We are also beginning to observe this process in Poland.

Among the economic benefits, the role of the reindustrialization of the economy in a new form is also emphasized, which is associated with job creation. This factor, however, raises more concerns in connection with the growing phenomenon of unemployment, as at least 10–20% of the existing jobs are at risk as a result of automation. Therefore, new social problems related to employment arise, as automation requires employees with higher qualifications, high specialization, and flexibility. New technologies are oriented toward greater opportunities for employees, not merely replacing them (*human-centered automation*), which requires us to redefine existing jobs and create completely new ones. First of all, the demand for engineering and IT staff is growing. However, concerns about rising unemployment are becoming real; there is even the specter of people fighting with robots for jobs, which has already happened in the case of manual work mechanization. These issues have already been raised in Poland by the Lewiatan entrepreneurs' organization (among others) in a document entitled "The future-of-work manifesto," postulating new employment models and a new order in the labor market thanks to developing the skills of professional adaptation to new conditions. The trends and postulates in this area are as follows:

- Adaptation of labor law and relationships as well as social security systems to new operating conditions of enterprises;
- Implementing lifelong learning;
- Extension of period of professional activity and protection of jobs by state;
- Enclusion of foreigners and those previously inactive to work;
- Better reconciliation of work with needs of private life.

In turn, the Warsaw Institute of Economic Studies conducted a study of the susceptibility to the automation of more than 900 professions in Poland and assessed that the Fourth Industrial Revolution could lead to the liquidation of 36% of jobs (mainly in the construction, food, and transport industries); the relatively safe professions represent only 28% of the market work. These percentages are much worse than the projected indicators for Benelux and Scandinavia, where the unemployment threat covers only about 20% of jobs.

The social effects of the Fourth Industrial Revolution promise to be far-reaching in terms of the changes in the life of society, although it is difficult to unequivocally assess these at present. We emphasized the need for changes in education and a greater diversity of the forms of employment and working time, which requires adaptation and organizational changes as well as a new psychological attitude. A similar issue will be the care of employee privacy and data security. The manipulation of supermarket customers and voters attempts to create a global cryptocurrency in blockchain technology (including Facebook), and expanding the surveillance system (especially the "Sharper Eyes" system in China) are other shortcomings. There are also other potential changes in social life. In Germany, the term "Leben 4.0" (Life 4.0) is even used. Although macroeconomic or even megaeconomic (global), all of these issues are also reflected in the management of a single company (however, to the extent resulting from the preparation of entrepreneurs for these challenges). This is evidenced by the examples cited in this chapter; however, the numerous effects already published in industry and daily magazines inform us about the beneficial effects and threats resulting from digitization.

Summarizing, the economic effects of the new stage of the industrial revolution are the result of the fast delivery of products according to the needs of individual customers; however, labor costs and economies of scale (the effects of extending the series) are less important, which is a new fact to be accepted in Poland. The introduction of robot taxation and universal basic income (regardless of employment) may prove to be certain remedies for unemployment threats, which have already been taken into account in several highly developed countries. We observe that, in addition to institutes and universities, some analyses and studies of these processes have also been undertaken by chambers of commerce and associations of family enterprises, usually in conjunction with audit and consulting companies.

Conclusion

The transition of the economy into servitization as a result of the fourth and fifth stages of the industrial revolution has brought about popularization of artificial intelligence devices and an increase in the role of the industry in structural changes. These new processes resonate with populist slogans; yet, they predominantly prove to be an important element of programs poised to combat unemployment and create more jobs (usually for employees with new qualifications). However, these processes have primarily become a decisive factor as well as the result of social and economic development (which scientists and politicians find quite interesting) and should also encourage entrepreneurs to make commercial use of the results of the scientific research. Generally, large corporations are aware of this fact, and some even decide to cooperate with other corporations with which they have previously been competing with the market (strategic alliances).

However, there are many concerns about the various and difficult-to-predict results of the massive spread of artificial intelligence. Nonetheless, small companies (especially family businesses) have been more cautious in this area despite their earlier experience. This demonstrates that the installation of robots in small companies results in technical progress, which proves decisive for the success on the market, improvement of working conditions for the work crew, and other beneficial social effects. *Personalization* of the production process and provision of services (i.e., a much better adaptation of products to meet customer needs) may prove to be especially beneficial. By the same token, making use of artificial intelligence devices in NGOs proves not only smart but also cost-effective in the long-run perspective.

References

Blung, M. (2000). *Teoria ekonomii. Ujęcie retrospektywne [Economic theory: A retrospective approach]*. PWN.

Council of Ministers. (2016). *Strategy for Responsible Development for the period up to 2020 (including the perspective up to 2030)*. Council of Ministers of the Republic of Poland.

Furmanek, W. (2018). The fifth industrial revolution. Explication of the concept. *Education—Technology—Computer Science*, *24*(2), 275–283.

Götz, M., & Gracel, J. (2017). The fourth-generation industry (industry 4.0)—challenges to research in the international context. *Vistula Scientific Quarterly*, *51*(1), 217–235.

Iwano, K., Kimura, Y., Takashima, Y., Bannai, S. & Yamada, N. (2016). *Future services & societal systems in society 5.0*. Center for Research and Development Strategy, Japan Science and Technology Agency.

Kiełczewski, D. (2012). On servitisation of economy. *Optimum. Economic Studies*, *58*(4), 37–44.

Laudante, E. (2017). Industry 4.0, innovation and design. A new approach for ergonomic analysis in manufacturing system. *The Design Journal*, *20*(1), 2724–2734.

Michałowski, B. (Ed.). (2018). Internet of things (IOT). In *Artificial intelligence (AI) in Poland*. Instytut Sobieskiego.

Moeuf, A., Pellerin, R., Lamouri, S., & Tamayo-Giraldo, S. (2017a). *Industry 4.0 and the SME: A technology-focused review of the empirical literature*. Paper presented at the 7th IESM Conference, Saarbrücken (Germany), October 11–13.

Moeuf, A., Pellerin, R., Lamouri, S., Tamayo-Giraldo, S., & Barbaray, R. (2017b). The industrial management of SMEs in the era of industry 4.0. *International Journal of Production Research*, *56*(3), 1118–1136.

Nahawandi, S. (2019). Industry 5.0—A human-centric solution. *Sustainability*, *16*(11).

Olender-Skorek, M. (2017). The fourth industrial revolution and some economic theories. *Social Inequalities and Economic Growth*, *51*(3), 38–49.

Sawhney, R., Treviño-Martinez, S., Macias de Anda, E., Tortorella, G. L., & Pourkhalili, O. (2020). A conceptual people-centric framework for sustainable operational excellence. *Open Journal of Business and Management*, *8*(3), 1034–1058.

Schwab, K. (2017). *The fourth industrial revolution, world economic forum 2016*. Penguin Random House.

Szász, L., Demeter, K., Rácz, B.-G., & Losonci, D. (2021). Industry 4.0: A review and analysis of contingency and performance effects. *Journal of Manufacturing Technology Management*, *32*(3), 667–694.

Sztuczna inteligencja [Artificial intelligence]. www.sztucznainteligencja.org.pl.

Tadeusiewicz, R. (2007). *Odkrywanie właściwości sieci neuronowych przy użyciu programów w języku C#. [Discovering Properties of Neural Networks Using C#]*. PAU.

Tobon Valencia, E., Lamouri, S., Pellerin, R., Dubois, P., & Moeuf, A. (2018). The integration of ERP and inter-intra organizational information systems: A literature review. *IFAC Papers on Line*, *51*(11), 1212–1217.

Tortorella, G. L., Mac Cawley Vergara, A., Miorando, R., & Sawhney, R. (2020). The role of industry 4.0 on the association between customers' and suppliers' involvement and performance improvement. In C. Machado & J. P. Davim (Eds.), *Industry 4.0. Challenges, trends, and solutions in management and engineering* (pp. 133–160). CRC Press.

Tortorella, G. L., Pradhan, N., Macias de Anda, E., Trevino Martinez, S., Sawhney, R., & Kumar, M. (2020). Designing lean value streams in the fourth industrial revolution era: Proposition of technology-integrated guidelines. *International Journal of Production Research*, *58*(16), 5020–5033.

Tworóg, J. & Mieczkowski, P. (2019). *A Short Tale of Society 5.0. How to live and function in the times of Industry 4.0 and 5G network*. Digital Poland Foundation.

Zaborovsky, V., Lukashin, A., & Muliukha, V. (2015). Robotic operations network: Cyber-physics framework and cloud-centric software architecture. In D. B. Rawat, J. J. P. C. Rodrigues, & I. Stojmenovic (Eds.), *Cyber-physical systems: From theory to practice* (pp. 259–282). CRC Press and Taylor & Francis Group.

Závadská, Z., & Závadský, J. (2018). Quality managers and their future technological expectations related to industry 4.0. *Total Quality Management & Business Excellence*, *31*(7–8), 717–741.

Index

www.ingramcontent.com/pod-product-compliance
Lightning Source LLC
Chambersburg PA
CBHW060250220326
41598CB00027B/4050

* 9 7 8 1 0 3 2 0 3 0 4 5 6 *